A Slim Book about Narrow Content

Gabriel M. A. Segal

A Bradford Book

The MIT Press

Cambridge, Massachusetts

London, England

This book was set in Sabon by Asco Typesetters, Hong Kong, and was printed and bound in the United States of America.

First printing, 2000.

Library of Congress Cataloging-in-Publication Data
Segal, Gabriel.
  A slim book about narrow content / Gabriel M. A. Segal.
      p.   cm. — (Contemporary philosophical monographs ; 1)
  Includes bibliographical references and index.
  ISBN 0-262-19431-7 (hardcover) — ISBN 0-262-69230-9 (pbk.)
  1. Supervenience (Philosophy) 2. Concepts. I. Title. II. Series.
B841.8.S44 2000
128'.2—dc21
                                                      99-054989

For Heidi

Contents

Acknowledgments

Many friends, acquaintances, and colleagues have contributed to this monograph. Mike Martin, David Papineau, Barry C. Smith, Scott Sturgeon, and three anonymous reviewers read at least one whole draft of the work. Ned Block, Dick Boyd, Peter Carruthers, Mat Carmody, Tim Crane, Tom Crowther, John Dupré, Frank Jackson, Joseph LaPorte, John Milton, Mark Sainsbury, and Rob Wilson all read some bits. Among them, they spotted many errors and infelicities and made lots of constructive comments and suggestions, for which I am most grateful.

I have wandered around, giving versions of chapters 2 and 3 in talk form. I am grateful to audiences at Cornell University, Amherst College, Birkbeck College at London, King's College at London and the University of Nottingham for discussions.

The work was completed while I was on a research leave funded by the Arts and Humanities Research Board of the British Academy. I am grateful for their support.

Thanks also to Mat Carmody for preparing the bibliography and index.

A Slim Book about Narrow Content

1 *Introduction*

Some properties are relational, some are not. What makes a property relational is that an object's possession of it depends not only on the object itself but also on circumstances external to it. Obvious examples include such things as being located in Lyons, being shorter than Muhammad Ali, and being a nephew. Less obvious examples derive from the theory of Special Relativity: having a particular mass or length or even shape turns out to be relative to a frame of reference. Being cubical, for example, a property that would intuitively seem to be paradigmatically nonrelational, in fact depends partly on something other than the possessor of the property: a cube is cubical not all by itself but relative to a frame of reference. Many such cases have been controversial in philosophy, including the ultimate nature of reality. At one extreme, relativists have argued that the nature of reality is largely dependent on perceiving and conceiving minds or cultures; at the other extreme, realists have argued that most of reality is as it is, quite independently of the human epistemic condition.

I will call properties that are not relational, "intrinsic." Intrinsic properties include such things as chemical and microstructural constitution. My wedding ring is composed of gold and platinum, with a certain amount of impurity. Its having this composition does not depend on anything outside the ring itself, and hence it is an intrinsic property of the ring. Similarly, my favorite coffee mug has the property of being composed of an arrangement of particular ceramic crystals. This property is intrinsic to the cup and involves nothing external to it.

Relational properties often have intrinsic counterparts. Corresponding to the relativistic property of mass, there is rest mass. Rest mass is the mass of a body when at rest, as it would be measured by an observer who is at rest in the same frame of reference. The rest mass of a body is its intrinsic contribution to its relativistic mass, relative to this or that frame of reference. Analogously for being hard or brittle or flexible: these properties might be understood relationally, in terms of being resistant to certain impacts, breakable by certain impacts, or prone to alter shape when subject to certain forces, respectively. But in each case there is an intrinsic counterpart, for example, an object's proneness to resist certain impacts depends on how its components (molecules or whatever) are bound together, as does an object's proneness to break or to alter shape.

It should be obvious that a good understanding of the nature of a property requires knowing whether it is relational or intrinsic. One would be in the dark as to the nature of avuncularity, for example, if one did not know the kinship relationships involved. Similarly, humans were somewhat in the dark as to the real nature of mass, shape, and so on, prior to Einstein's discovery of Special Relativity. Equally, one would be somewhat in the dark about the property of being made of gold if one did not know that it was intrinsic, a matter of molecular constitution, rather than, say, etiology or market value.

The concern of this book is whether certain psychological properties are relational or intrinsic. Specifically, it will be concerned with what might called the "cognitive properties" or "cognitive content" of psychological states. By "cognitive properties" I mean those properties that account for the role of these states in typical psychological predictions and explanations. Suppose, for example, that Yogi believes that orangutans are omnivorous, and that all omnivores like chocolate. We might then predict that, if he considers the matter, he will come to believe that orangutans like chocolate. Or, after the fact, we might explain his believing that orangutans like chocolate by citing the other two beliefs. Similarly, if Yogi himself wants to buy some chocolate and believes that in order to buy some chocolate, it is necessary to go the shops, then we might predict that he will go the shops. Again after the fact, we might explain why he went to the shops in terms of his desire and his belief.

Psychological explanations of this sort evidently draw on a specific range of properties of the states they cite. The properties appear to be specified by the embedded complement clauses of propositional-attitude attributions, the "that" clauses of "believes that p," "doubts that q," "hopes that r," etc. These clauses give the contents of states they ascribe, in the sense that they specify what is believed, doubted, hoped, and so on.

The term "content" used in this loose, intuitive sense, is rather vague and ambiguous. Suppose, for example, that each day Abraracourcix believes that the sky will fall on his head tomorrow. Does he have the same belief, that is, a belief with the same content, from day to day? In one sense, it appears not, since Monday's belief has different truth conditions from Tuesday's. He does not, then, retain a belief with the same truth conditions. But it is tempting also to suppose that in another sense of content, his belief each day has the same content: that the sky will fall on his head tomorrow. A related but different ambiguity in "content" occurs with examples like believing that water boils at 100 degrees Centigrade and believing that H_2O boils at 100 degrees Centigrade. These appear to have the same referential content: the beliefs predicate the same concept (boiling at 100 degrees C) of the same substance (water/H_2O). But they appear to differ in cognitive content, since the beliefs would play different roles in a person's thinking.

From now on when I use "content" unmodified, I mean "cognitive content" rather than, say, truth-conditional or referential content. I will not make any initial assumptions about the relationship among different notions of content, about whether two or more notions might pick out the same phenomenon, and so on. Nor will I make any initial assumptions about the precise relation between cognitive content and the complement clauses of attitude attributions. I leave it open, for now, whether identity and distinctness of complement clauses correspond directly to sameness and difference of content. These matters will come up for explicit discussion as we proceed.

I will merely assume that there is such a thing as cognitive content, that it drives standard psychological explanations, and that we use attitude attributions to get at it in some manner or other. I will use "content" to refer to properties and items as they would be individuated in a true psychological theory. So questions about sameness or distinctness of contents are questions about the taxonomy of a correct psychological theory. I will also assume, except where I explicitly say otherwise, that psychological items, states and events, at least cognitive and representational ones, are to be individuated by their contents. So questions about sameness and difference of beliefs, concepts, etc., are questions about sameness and difference of contents.

Although I will focus almost exclusively on ordinary commonsense psychology and the propositional attitudes that are its main concern, I intend all the main arguments that I offer to extend to any branch of scientific psychology that recognizes contentful states and to all such states, including perceptual states, states of the Freudian unconscious, tacit cognition of language, neonate cognition, animal cognition, and so on.

1.3 TWIN WORLDS

Are cognitive contents relational or intrinsic? Suppose, for example, that Zowie believes that her engagement ring is studded with diamonds. Does her having a belief with that content essentially involve any relations to anything beyond Zowie herself? The easiest way to get a handle on this question is to consider in what kinds of environments it would be possible to have a belief with just that content. Could Zowie have such a belief in a world in which her engagement ring did not exist? (Poor deluded Zowie, driven insane by love unrequited.) Could she have such a belief in a world in which diamonds did not exist? Or in a world with no other humans? (Zowie, an artificially constructed brain in an extraterrestrial scientist's laboratory.)

A particular kind of thought experiment introduced by Hilary Putnam (1975a) is very useful for rendering such questions vivid. Thought experiments of this kind involve imagining or conceiving of what we can call "twin" subjects in "twin" worlds. Twin subjects are microstructural duplicates of each other: they are structurally identical in respect of the elementary particles, the atoms and molecules, the nerve cells and their interconnections, the neural structures, etc., that make them up. Twin worlds are also microstructural duplicates, except in one specially selected respect. The thought experiment involves assessing whether the difference between the worlds entails a difference in the psychological properties of the subjects.

We have our Earthly subject, Zowie, believing that her ring is studded with diamonds. Let us suppose that Zowie lived in the seventeenth century, prior to the discovery that diamonds are made of pure carbon. We now imagine a twin Zowie on a twin Earth. Twin Zowie is a microstructural duplicate of Zowie. At any given moment, her brain, central nervous system, and everything else within her body are in exactly the same configurations as Zowie's. Twin Earth is exactly like Earth, except in respect of diamonds. On Twin Earth there are stones that are just like diamonds in all superficial respects: they are very hard, when of good quality they sparkle enticingly in the light, etc. They are so like diamonds that nobody on Earth or Twin Earth could have distinguished the two in the seventeenth century. These stones are regarded on Twin Earth just as diamonds are on Earth. Twin Earth counterparts of English speakers call them "diamonds," but we can call them "twin diamonds." Twin diamonds differ from Earth diamonds in their internal constitution, being made up not of carbon but of a kind of aluminum oxide. If a modern day geologist found a twin diamond, she could distinguish it from a normal Earth diamond in her laboratory, and she would probably classify it not as a type of diamond but more likely as a type of sapphire.

Zowie and Twin Zowie both say "My engagement ring is studded with diamonds." Are the concepts expressed by their words "diamond" the same? Or are they different?

One might be tempted to answer that they are different, by some reasoning like the following. What Zowie is saying is that her engagement ring is studded with diamonds. If the stones are indeed diamonds, what she says is true. And if what she says is what she believes, then what she believes is true too. Had the stone on her ring been a twin diamond, aluminum oxide rather than carbon, she would have said and believed something false. For twin diamonds are not diamonds: they just look like them.

Of course, Zowie does not know that diamonds are made of carbon. So she doesn't know that if her jewel is not made of carbon, then it is not a diamond. But that makes no difference. Many contemporary English speakers don't know that diamonds are made of carbon. Yet when they say "diamond" they still mean *diamond*. If one of them pointed to a twin diamond and said, "That's a diamond," he would be saying something false. If he believed what he said, he would have a false belief as well. And so it is with Zowie.

With Twin Zowie, things are reversed. When she uses the word "diamond," she doesn't mean *diamond*. She has never seen a diamond, nor has she met anyone who has seen one. In fact, she has had no contact with diamonds at all, no matter how indirect. When she says "diamond," she is using it to refer to what Twin English speakers normally refer to when they use the same word: twin diamonds. If her jewels are genuine twin diamonds, then what she says is true. And if she believes what she says, what she believes is true as well. What she believes is something we might approximately express by saying that her engagement ring is studded with twin diamonds. That is certainly different from what Zowie believes, so their beliefs have different contents.

Zowie and Twin Zowie are identical in all intrinsic respects. They differ only in their relationship to diamonds. So if the above reasoning is along the right lines, it shows that cognitive content depends partly on factors external to their subjects and so are partly relational. However, one might be tempted to doubt the conclusion and hold that the contents of the twins' beliefs are the same. After all, it is hard to see how the difference between diamonds and twin diamonds, a difference of which the Zowies are quite unaware, can make any difference to how the world appears to them or to how they think and reason.

Throughout this book, I will discuss arguments on both sides arising from this kind of Twin Earth experiment. For now, I will use it as a way to introduce some of the main ideas that will feature as we proceed.

1.4 SUPERVENIENCE

Twin-world experiments are fundamentally about supervenience. Philosophers have refined a number of useful senses of "supervenience" (see Kim 1984 for discussion). But I will just stick with a simple and rough one: a set of properties B supervenes on a set of properties A if and only if (iff) any two objects identical in respect of A properties must be identical in respect of B properties too. Weight, for example, supervenes on mass and local gravity: any two objects of the same mass, subject to the same local gravity, must have the same weight. Weight does not supervene on size, however, since two objects of the same size may have different weights. If the twin Zowies differ in respect of the contents of their beliefs, then these contents fail to supervene on intrinsic properties. (Remember that, by hypothesis, the twins have identical intrinsic, microstructural properties.) By contrast, if any possible twin of Zowie, no matter what her external environment is like, must share all her cognitive contents with Zowie, then cognitive contents do supervene on intrinsic, micro-structural properties, or to use a common abbreviation, they are "locally supervenient."

The question of whether content is locally super-venient is not quite the same as whether it is intrinsic to the subject. There are at least two reasons for this. The first is that putting the issue in terms of local supervenience on microstructure leaves no comfortable place for a Car-tesian dualist to enter the discussion. In effect, it assumes that the subject of cognitive contents—the object to which cognitive properties are or are not intrinsic—is a physical thing, in the minimal sense of being made out of atoms, molecules, and so on. This is not an assumption the Cartesian would share. A Cartesian who believed in the intrinsicness of content would not need to hold that Zowie and Twin Zowie are cognitively exactly similar: whether they are is a question about their immaterial souls, not their material brains and bodies. The Cartesian might or might not believe in the local supervenience of the mental on the physical. But that would not bear directly on the question of whether mental properties are intrinsic prop-erties of their immaterial subjects. (It is often claimed that Descartes himself believed that the mental is intrinsic; whether he did or not is a question that I refrain from addressing.)

The second reason why the questions of local super-venience and intrinsicness come apart is that a property might be relational, at least in a weak sense, and yet be locally supervenient. This is rather obvious if one thinks abstractly: there does not seem to be any principled reason why some relational property R should not be locally supervenient. That would just mean that any twins would necessarily be identical in respect of R: they would either all have it or all lack it. As an illustration, not to be taken too seriously, suppose that height supervenes on micro structure. Then any twins are the same height. But having a given height automatically puts one into a relation with a given number: being six feet tall, for instance, puts one in a relation with the number six, specifically the relation "is ____ feet tall." Being six feet tall is thus locally supervenient—it is shared by all twins—but also, in this weak sense, relational.

Someone might reasonably hold that content is indeed both locally supervenient and relational if they held that contents are relations to abstract objects, such as properties. One might think, for example, that thoughts about diamonds involve relations to the property of being a diamond, where a property is an abstract object that exists independently of its instances. Then any being thinking about diamonds will stand in a relation to this property, even if there are no diamonds in its environment. Colin McGinn (1989) calls this position "weak externalism." Whether weak externalism is true is an interesting question, but is irrelevant to the topic of this book.

This book is concerned with whether content essentially involves relations to external, concrete, contingently existing things. The Twin Zowie story illustrated the idea that some sort of relation to samples of the kind of thing a concept represents is necessary for possession of the concept. The story raised the question of whether a relationship with diamonds is necessary for having a concept of diamonds. And many have thought that this is so, that concepts of natural kinds, like diamonds, do require some real relationship with actual instances. It has also been argued, by Tyler Burge (e.g. 1979) in particular, that certain relations to other language users are determinants of content. For these sorts of items—the kinds that concepts represent and other language users—being intrinsic and being locally supervenient coincide.

For if content is locally supervenient, then it will always possible to conceive of a possible environment in which someone has a state with the relevant content but in which the items do not exist. Take subject Z in a certain representational state S that represents some kind K. We can always imagine a microstructural duplicate of Z in an alien scientist's vat and in a world devoid of diamonds (trees, water, tigers, or whatever) and devoid also of other speakers. Or if some glitch comes up with that kind of example, we can imagine that the twin arises as the result of a quantum accident: he or she suddenly emerges in outer space and survives for a short while, floating in the void. If S is locally supervenient, then the twin would be in state S. And this would entail that S is intrinsic, rather than relational, for in the twin's environment there are no Ks for it to be related to.

My aim is to argue for the local supervenience of content. Given what I have just been saying, and given the minimally materialist assumption that bearers of cognitive properties—humans, animals, cognitive systems of all kinds—are made up out of elementary particles, arguing for local supervenience is a way of arguing that content is intrinsic. The position I will be defending is a version of what is called "internalism" or "individualism," and I will use both labels to refer to it. However, both labels are vague and should be taken to gesture towards a family of positions rather than any very specific thesis.

The thesis with which I will be mainly concerned is then this: being in a state with a specific cognitive content does not essentially involve standing in any real relation to anything external. Cognitive content is fully determined by intrinsic, microstructural properties: duplicate a subject in respect of those properties and you thereby duplicate their cognitive contents too.

Internalism, the thesis that content supervenes on microstructure, thus embodies two ideas. The first is that content is not relational, does not depend anything outside the subject. The second is that it does depend on microstructure. The latter idea is not without substance, since an alternative would be that content does not depend on anything. Zowie believes that diamonds sparkle. Suppose now that Zowie has a twin who is identical both in microstructure and in relational properties. Perhaps this twin Zowie does not believe that diamonds sparkle. If this were possible, having that belief could still be intrinsic to Zowie; it is just that it would not depend on certain other of her intrinsic properties, the microstructural ones. This is a possibility that some courageous people might accept (see, for example, Crane and Mellor 1990 and Cartwright 1994). This courageous position is closely related to the Cartesian one, although it can allow that psychological properties are properties of bodies or brains. A consequence of my arguments for local supervenience will be that this position should be rejected.

The internalism I will argue for likens content to properties like being hard, being brittle, or being liquid, in their intrinsic versions. One can explain a thing's possession of these properties in terms of properties of and relations among its constituent parts. A diamond is hard because of the way its crystals are bound together. A ceramic cup is brittle because of irregularities at the boundaries of the crystals that make it up. The water in a glass is liquid because of the way its molecules are loosely bound together. This sort of explanation, which could be called "systematic" (loosely to follow the usage of Haugeland 1978), seems to be fairly widespread. It applies to functional properties like the ones we've just considered, along with being transparent, being plastic, and so on.

Systematic explanation is the norm when it comes to explaining how artefacts work: when one explains how a car or an espresso machine or a dishwasher works, one cites what its component parts are, what each one does, and how the combined actions of the various parts suffice for the machine to do what it is supposed to do (see Haugeland 1978 and Cummins 1983 for more on this). It applies in the explanation of many different kinds of natural properties. For example, thermodynamic properties of gases are explained in terms of properties of and relations among the component molecules. And biomedical properties of hearts, lungs, and so on, are explained in terms of properties of and relations among their parts (the auricle, ventricle, etc.).

The idea, then, is that cognitive properties, like so many others, can be given systematic explanations in terms of properties of and relations among their bearers' parts. I have largely resisted the temptation to say "physical" parts, since I find that term unhelpful at best. The word "physical" tends to be used by those—physicalists and dualists alike—who think that subject matters of intellectual inquiry divide in some principled way into the physical, the not-yet-shown-to-be-physical, and (possibly) the nonphysical. This is wrong.

I do not know which properties and relations of which parts are the relevant ones for explaining cognitive properties. They might be functional properties of and relations among neurons, of a sort within the descriptive reach of current neurology. They might be computational properties of neurons, of a sort within the reach of current computer science. Or they might be—and I suspect they are—as yet undreamed of properties of some kind of neural wotsits to be discovered by some future science that develops at the overlap of neurology and psychology.

The advantage of focusing on microstructure, that is, on the level of elementary particles, is that whatever it is that determines content probably supervenes on it. Fix an object's microstructure and you fix its atomic and molecular structure, its neurological and computational properties, and so on. Or so I will assume, anyway, to facilitate exposition.

1.6 THE HARDNESS OF THE SUPERVENIENT "MUST"

The local supervenience thesis—if two beings are identical in respect of their microstructural properties, then they must be identical in respect of their cognitive contents— can be interpreted in different ways, depending on the strength of the "must." The point calls for discussion.

Internalism is often held to involve a notion of "metaphysical necessity." Metaphysical necessity was first described clearly by Saul Kripke (1971, 1980). For illustration, consider the (true) identity statement that Hesperus = Phosphorus ("Hesperus" and "Phosphorus" being two names for the planet Venus). It is metaphysically necessary that Hesperus = Phosphorus. It is not logically necessary, since no amount of purely logical deduction could reveal its truth. It is not epistemically necessary, since someone might perfectly well not know that Hesperus is Phosphorus (the ancient Greeks didn't). However, it is necessary in a very strong sense: it could not have been otherwise, no matter what. No matter how different the world might have been in respect of the laws of nature or anything else, Hesperus could not have been a different object from Phosphorus. Since Hesperus is the very same object as Phosphorus, it could not possibly have been some distinct object, for then it would have been distinct from itself, which is clearly impossible.

Identity statements involving kind terms and certain sorts of statements about a kind's constitution have also been seen as metaphysically necessary. For example, some would hold that it is metaphysically necessary that diamonds are made of carbon. Try to conceive of a possible world in which diamonds are not carbon, and you will fail. You can conceive of a world with twin diamonds, stones that resemble diamonds but have a different molecular constitution, but twin diamonds are not diamonds. So some would argue (Kripke 1980).

An internalist who held that microstructural duplicates must, as a matter of metaphysical necessity, be cognitive duplicates, would hold that all possible twins are in the same psychological states. If Zowie believes that orangutans are omnivorous, then there is no possible world, however different from the actual one, inhabited by a twin Zowie who doesn't share the belief. As I said, internalism is indeed sometimes said to involve a notion of metaphysical necessity. I do not wish to defend quite such a strong thesis.

The main reason for this is that I have a worry about the methodology of assessing claims for metaphysical necessity. The standard way to find out whether a given proposition is metaphysically necessary is to try to conceive of a possible situation in which it is false. But it is not obvious why there should be any very strong connection between what we can conceive of, or imagine, and truths about the world. It is not obvious that if one cannot conceive of something's being the case, it follows that it really could not be the case. Whether the entailment goes through depends at least in part upon what accounts for our capacities to conceive, on how our conceptual faculties function. And this is not a subject about which much is known.

Further, one should doubt that we always know what we can and cannot conceive. There are those who would claim that they can conceive of diamonds not being made of carbon. Kripke might argue that these people are confused about what they can conceive. Perhaps that is correct. But that would just show that we are not infallible about the content of our own conceptions. Of course, it remains possible that we are reasonably reliable judges of our own conceptions, and that there are cases where our judgements about them can be trusted. But this optimism may be premature. Once again, until we have a better theory of our conceptual faculties, of how they work, we should be cautious about our judgements.

Putting the two points together reveals a gap between what we think we can conceive and what is objectively possible or impossible. There is no sure route from the former to the latter. Since conceivability is the chief method of assessing claims of metaphysical necessity, I think such claims are incautious. Hence it seems that it would be incautious for an internalist to make any claims about metaphysically necessary supervenience.

I will assume that the question of whether a property is relational or intrinsic is a question about natural necessity, or laws of nature. If it is nomologically possible for twins to have different contents, then contents are relational and not fully explicable in terms of microstructure. The thesis that nomologically possible twins must have the same contents is an interesting internalist thesis in itself, and one worth arguing for. For it shows us what content is actually like, no matter what the metaphysical possibilities are.

Here is an analogy. As I mentioned above, according to the theory of Special Relativity, properties like mass, size, and shape turn out to be relative to frames of reference. Special Relativity is an empirical theory that tells us about the real natures of mass, size, and shape. Nevertheless, it seems conceivable that Special Relativity might have been false. It seems as though one can conceive of a world in which the theory does not apply. But surely this apparent conceptual possibility tells us nothing interesting about what mass, size, and shape are actually like.

There are three reactions we might have to this thought experiment. We might hold that it tells that, e.g., mass might have been intrinsic, although it is actually relational. The relationality of a property would then be not essential to it. Or we might hold that the thought experiment tells us nothing about mass, but reveals the metaphysical possibility of a nomologically impossible world in which objects don't have mass but have instead some counterpart property, shmass. Shmass might be rather like mass, but it wouldn't be the genuine article. Or we might hold that our thought experiment only presents a world in which Special Relativity would seem to be false, a world in which we would be misled about the laws of nature and the nature of mass. (See Shoemaker 1998 for discussion.)

On all three views, the interesting truth about mass—the fact that it is relational—relates to the natural laws that govern it. The second and third views might allow that the relationality of mass is also metaphysically necessary. But this would only be because the metaphysical necessities flow from the natural, nomological ones. Hence on all three views, the focus of interest would be on natural laws and empirical theories.

So, from now on, I will standardly use the various notions that involve modals like "must" in line with the above, to invoke natural or nomological necessity. Thus local supervenience is the thesis that microstructure nomologically determines cognitive properties, that twins are nomologically possible twins, and so on.

Content that is locally supervenient is often called "narrow"; content that isn't is called "wide" or "broad." I will argue that narrow content is genuinely representational, honestly semantic. A number of internalists have argued for varieties of narrow content that are not directly representational but relate to representation proper only indirectly. For example, Jerry Fodor (1987) once argued that narrow content is a function from contexts to broad contents. To illustrate, consider Zowie once again. On that position (which Fodor no longer holds; see Fodor 1994) it is conceded that Zowie and Twin Zowie have thoughts with different semantics: one being about diamonds, the other about twin diamonds. So they have thoughts with different broad contents. But the thoughts also share a narrow content in virtue of the following: if Zowie had been in Twin Zowie's context, her thought would have been about twin diamonds, and if Twin Zowie had been in Zowie's context, her thought would have been about diamonds. And indeed, for any context C, Zowie and Twin Zowie would have thoughts with the same broad contents had they been in C. Thus the thoughts instantiate the same function from contexts to broad contents, and this is what gives the same narrow content.

Others have argued for a "functional role" theory of narrow content. A state's functional role is given by its causal potentialities: roughly, what would cause it to occur, and what, in turn, it would cause to occur, in various possible circumstances. So two states have the same functional role if their potential role in the causal nexus is the same. Some functional-role theorists hold, as Fodor did, that the representational contents of twins' states are different, and hence broad. But they construct a notion of narrow content in terms of functional role and hold that twins' states have the same narrow content because they have the same functional role.

Although I sympathize with much of the motivation for these versions of internalism, I wish to defend a different one. The version I will defend holds that narrow content is a variety of ordinary representation. Narrow content is just content, to be understood in roughly the terms it always has been understood (at least since Frege), such terms as sense, reference, truth, extension, "satisfaction" in the technical, Tarskian sense, "aboutness" in the philosophers' sense, and "intentionality" in roughly Brentano's sense. My view mandates denying that Zowie and Twin Zowie's "diamond" concepts have different extensions. In fact, my view is that both Zowies' diamond concepts apply to both diamonds and twin diamonds, so contrary to what some might initially think, if Zowie pointed to a twin diamond and said "That's a diamond," she would be saying something true in her idiolect.

I will not, however, argue that all representational content is narrow. It is necessary to make an exception for singular, demonstrative concepts such as those expressed by "this ring." The basic idea is simple. Suppose that Zowie and Twin Zowie have exactly similar engagement rings. Each twin points to her ring and thinks *This ring is beautiful*. It is clear that the referents of the demonstrative concepts are different. For reasons I will come to in chapter four, this leads to a legitimate notion of wide content. But it is arguable that this sort of wide content is not properly classed as psychological or cognitive.

I will not offer any "philosophical theory" of content. Trying to develop a philosophical theory of content is rather like trying to develop a philosophical theory of heat or water. Content is a real natural phenomenon. The most we should hope for by way of a "philosophical" theory of content is one that will tell us what content *is*, where the "is" has roughly the force it has in "Heat is the motion of particles" or "Gold is the element with atomic number 79." It may be that we are even asking too much when we ask for a theory of content in this sense. Maybe there is no answer to the question "What is content?" After all it is pretty unclear how to give a general account of heat that applies across solids, plasmas, and vacuums.

Be all that as it may, there is certainly a great deal we can discover about the nature of content and about how the organization of matter gives rise to it. The point is that we should not expect to discover too much from the armchair. Discovering the true nature of content should be a scientific enterprise (whether we also call it "philosophical" or not). These enterprises usually progress slowly because they involve a great deal of empirical and technical work. It might take many decades of detailed research before we make real progress. So a defense of internalism should not require defense of a theory of content. Proposed theories may well be premature. It should suffice to cast reasonable doubt on externalism, to motivate internalism, and to provide reasons to believe that good psychology is, or could be, internalist.

1.8 TACTICS

I do not claim to have any convincing, knockdown argument for internalism. I do not claim to have conclusive arguments against all varieties of externalism. But I think I can make a decent case for internalism. The basic strategy is to undermine what I think are the most popular and influential externalist theses and to show that an internalist alternative is workable and attractive.

What follows comes in four chapters. The first two offer arguments against two leading externalist theses. In chapter 2, I address externalism about natural-kind concepts and present an argument against it based on the existence of empty kind concepts, ones that lack an extension. In chapter 3, I present an argument against a popular form of "social externalism," the view that the content of many concepts depends in part on the views of experts, lexicographers, and so on. The message of the first two chapters will be that all general (nonsingular) concepts have a narrow content. The considerations leave open whether general concepts also have broad contents. Chapter 4 considers and rejects the leading two-factor theories that endorse both broad and narrow content for general concepts. Chapter 5 outlines and defends a radical alternative version of internalism, arguing that extension conditions are narrow.

World Dependence and Empty Concepts

Water and twater

Here is the original Twin Earth experiment from Putnam 1975a. We consider a time prior to modern chemistry, say 1750. We consider a typical Earthling, Oscar. We consider a Twin Earth that is just like Earth except that where Earth has water, Twin Earth has twater, a substance with a chemical composition different from Earth water, which we can abbreviate "XYZ." XYZ looks, feels, and tastes like water and is similar to it in observable macroscopic respects. But it is not H_2O. On Twin Earth there is an exact duplicate of Oscar, Twin Oscar. Now suppose that both Oscar and Twin Oscar say something using the word (or word form) "water," e.g., "Water is thirst-quenching." Do they mean the same thing by their words?

Putnam argued "No." Water, scientists tell us, is H_2O. XYZ is not H_2O. So XYZ is not water. Our word "water" is not true of XYZ. When Oscar used the word "water," he referred to the same substance as we do now. So Oscar's word "water" referred only to H_2O. But Twin Oscar's word "water" of course referred to the substance in his local environment, which is not H_2O, and hence is not water. So Oscar and Twin Oscar referred to different kinds of substance. And, assuming that a difference in extension entails a difference in meaning, what Oscar and Twin Oscar meant was different as well.

Thus far we have an externalist conclusion about the meaning of a word. But it is a short step to a similar conclusion about the contents of the twins' psychological states; their beliefs, desires, and so on. Sentences express psychological states. What the twins say is what they mean, i.e., the content of the belief they express is just the content of the sentence they utter. So what the twins believe is different too. (Putnam [1975a] did not pursue the conclusion about psychological states, but others, such as Colin McGinn [1982] and Tyler Burge [1982] did, although not always by the short step just described.)

Putnam's Twin Earth example has become a sort of paradigm in the philosophies of language and mind. It has featured at the center of many discussions and has carried considerable influence. However, if one reflects on the details, it becomes clear that there are a couple of flaws. As presented, the example simply does not work. These flaws in themselves are not of great significance and can be remedied simply by switching examples. But details can matter. Different examples work differently: they can raise different issues and generate different intuitions. For this reason it is worth taking the time to pick nits.

The example is supposed to show that intrinsically identical twins could have psychological states with different contents. The first problem is that humans are largely made out of H_2O. So if there is no H_2O on Twin Earth, then Twin Oscar isn't Oscar's twin. So the example falters near the start. This point has often been noticed and clearly is not of enormous significance. But it is not totally without ramifications. Since it would be awkward for externalism to rest on an argument that began, "If humans were not made out of water, then ...," it would be more expedient for the externalist to look for a different example.

The second problem is more interesting. This is the point made by Thomas Kuhn that nothing with a significantly different structure from H_2O could be macroscopically very like water.[1] If Kuhn is right about this, it follows that the example is nomologically impossible. This is important. Suppose we take it that Twin Earth is conceptually possible but nomologically impossible. And suppose that we have the externalist intuition that in such a case Oscar and Twin Oscar would mean different things by their words. This would show that externalism accords with some of our intuitions. As far as we can tell by intuition alone, it seems that twins might have psychological states with different contents. But it does not tell us about the actual laws that govern content. And this is one of the things we should care about. (See Fodor 1994 for discussion.)

Again, the conclusion is that we need to consider further examples. For if we found again and again that putative examples were nomologically impossible, then the internalist should be encouraged. Why would they all be nomologically impossible? Probably because there is something fundamental about how content fits into the natural world that constrains it to be narrow. And that would suggest that content is really intrinsic, not just relational and locally supervenient for adventitious reasons. It would suggest, in other words, that according to (some of) our intuitions, content is relational, but that those intuitions are wrong. All this depends on how other examples go, of course. So let us consider one.

Topaz and citrine

Topaz and citrine are attractive gemstones, usually of a yellow or amber color. Topaz is precious, and citrine semiprecious. They are different kinds of stone, in the sense of being composed of different chemical compounds. Topaz is $Al_2SiO_4(OH, F)_2$ and citrine, a type of quartz, is SiO_2. Typical samples of yellow topaz and citrine are indistinguishable to the eye and other unaided senses. But they are easily distinguished by their different refraction indices.

Let us now consider two twin Earths. On TE_1 there is only topaz and no citrine, on TE_2 there is only citrine and no topaz. Roll the clock back again to 1750, when no one on the planets could have told the difference. Suppose that on both planets, speakers use "topaz" to talk about the stones. Now, the externalist version of the story runs more or less as before. When the technology and science develops, scientists investigate the stones. TE_1 scientists say, "Ah, topaz is $Al_2SiO_4(OH, F)_2$." TE_2 scientists say, "Ah, topaz is SiO_2." If, now, a TE_1 speaker encountered a citrine (imported, say, from TE_2) and called it a topaz, he would have spoken falsely: "topaz" in his mouth refers to topaz, $Al_2SiO_4(OH, F)_2$. But "topaz" in a TE_2 speaker's mouth refers to citrine, SiO_2. Given that the meaning of "topaz" did not change with the discovery of the stones' chemical composition, we can infer that even before the scientific facts were discovered, the words and corresponding concepts of TE_1 and TE_2 speakers also differed.

This example does not suffer from the flaws of Putnam's. Humans are not made out of topaz or citrine. And the example is nomologically possible.

The debate

Putnam's conclusions in "The Meaning of 'Meaning'" were moderate in three important respects. First, as noted, he argued only about the meaning of words and not about the contents of psychological states. Second, he confined his conclusions to natural-kind terms and did not extend them to other sorts of general terms. Third, he suggested that meaning could be decomposed into two factors. One factor consists of the extension conditions: Oscar's "water" is true of something iff it is a sample of H_2O, while Twin Oscar's "water" is true of something iff it is a sample of XYZ. The other factor is a "stereotype." This is, roughly speaking, a set of descriptions used to identify samples of the extension: for water, colorless, odorless, tasteless, etc. These descriptions do not provide necessary or sufficient conditions for something to fall under the term. They are merely heuristic markers. The stereotypes associated with "water" would be the same across twins. Their terms differ only in their extension conditions.

Many philosophers have made stronger claims than Putnam in all three respects. Burge, in particular, developed different kinds of examples designed to extend externalist conclusions to a far wider range of terms and concepts. These will be discussed in chapter 3. This chapter will be concerned only with natural-kind terms and concepts.

Many philosophers have also rejected Putnam's suggestion that word meaning (or representational content) could be neatly decomposed into two factors. They deny that it is possible to separate the relationally determined aspects of contents, extension conditions, from some more internal factor, such as a stereotype (see Burge 1982, 103).

We can distinguish three broad positions in the debate. Each of these comes in many varieties, so the real picture is complex.

There is what we might call "radical internalism." This is the thesis that any content relevant to psychology is narrow. The extreme version of radical internalism dispenses with the notions of extension and extension conditions altogether.[2] The less extreme version does not dispense with these notions but holds that extension conditions are themselves narrow. The claim is not, of course, that all samples of the extension of a concept are inside people's heads. It means that the extension conditions of a thinker's concepts are determined by intrinsic features of the thinker. For example, my concept of water has approximately the following extension condition: (x) ($water$ applies to x iff x is a sample of H_2O). The claim is that this feature of my concept is fully determined by my intrinsic properties. Any twin of mine in any environment would have a concept with the same extension condition.

The nonradical, middle position endorses a two-factor theory. It accepts the Putnam-inspired claim that at least some concepts have broad contents. But it also holds that every concept has a further, extension-independent content. A two-factor theory is rather widely held for definite descriptions. For example, "the present Queen of Denmark" actually denotes Queen Margaret. Counterfactually, if Margaret had had an elder sister, Ingrid, then "the present Queen of Denmark" would have denoted Ingrid. It is natural to think that the definite description would have had the same meaning and expressed the same concept in either case. Thus we can distinguish two factors that make up the full representational content of the definite description: the object-independent content, shared across the two cases, and the extension, or denotation, which differs across the two cases.

The key point about two-factor theories in general is that they allow for a kind of content that is extension-independent, in the sense that it does not essentially depend on the underlying nature of actual samples of the extension in the environment of the thinker. Putnam's claim was that the extension conditions of "water" are extension-dependent in this sense. Oscar's words and Twin Oscar's words differ in extension conditions because the underlying nature of samples of the extensions in the two environments differ. A two-factor theory, like Putnam's own, posits an additional level or kind of content that is extension-independent.

Not all two-factor theories make concessions towards internalism. An externalist might hold a two-factor theory and argue that both factors are partly relational. She might agree, for example, that the content of "the present Queen of Denmark" decomposes into a queen-independent, purely descriptive component and an extension (an actual queen). But she might argue that the content of the descriptive component itself depends on other external factors: perhaps the component concepts *queen* and *Denmark* are world-dependent in some way. Equally, an externalist might allow that there is an extension-independent content shared by Oscar's concepts and Twin Oscar's concepts but hold that this shared content depends on affairs outside the subjects' skins. It might, for example, depend on the existence of other people.

The concessive middle position is, however, internalist about the extension-independent content. According to this position, extension-independent content is narrow. So, for example, Oscar's and Twin Oscar's "water" concepts have a shared narrow content. This concessive two-factor theory is harder to identify than one might think. For the narrow factor is supposed to be distinct from extension conditions. But extension conditions are the paradigm of content, at least according to a certain outlook. There is therefore a question as to which kinds of narrow factor should be classified as content and which should not.

The radical externalist position rejects the other two positions. It holds that at least some concepts have only broad contents. There are two main versions of radical externalism. The extreme version simply identifies broad content with extension.[3] The less extreme version does not. It allows for something like sense. It allows that coextensive concepts can differ in content.[4] But it denies that what distinguishes coextensive concepts is narrow content. What distinguishes coextensive contents might itself be relational. Or, on a more standard view, what distinguishes coextensive contents is not a separable factor at all. Perhaps one could think of it as how the subject is related to the extension, or how the extension is presented to the subject in thought. On this view, these matters cannot specified in abstraction from the extension, but rather essentially involve it.[5]

I call the radical externalist position, as applied to natural-kind concepts, "the thesis of world dependence of kind concepts" (TWD). The purpose of this chapter is to refute TWD. I will argue that kind concepts have an extension-independent cognitive content. The conclusion will leave open the choice between two-factor theories and radical internalism. I will return to the outstanding choices in chapters 4 and 5. The overall aim is gradually to eliminate leading externalist options and then to motivate a version of radical internalism.

2.2   AGAINST TWD

TWD involves three characteristics. First, it means that the extension conditions of a nonempty kind concept depend in part on a real relationship between thinkers and samples in the external world. So thinkers must either have interacted with samples or known someone who has interacted with samples or been in some other form of direct or indirect causal contact with the samples.[6] Second, it means that extension conditions are essential to the cognitive content of the concept. So if concept C is world-dependent, then any concept that has different extension conditions from C has a different cognitive content from C. The third characteristic of world dependence rules out decomposition of content into two factors.

I will argue that TWD cannot account for the existence of empty kind terms and concepts, kind terms and concepts that do not refer to anything.

Argument sketch

Empty terms and concepts provide the largest problem for the thesis of world dependence. One way of getting at the difficulty is to consider the fate of a kind term in the counterfactual circumstance in which it has no extension. Suppose that there were no water. (The supposition may require a considerable stretch of the imagination, but let us do our best). Suppose that there were no aluminum or topaz or quarks or polio. But imagine that otherwise things are as much like they are as they could be, compatible with this counterfactual premise. Then what would happen to the meanings of the terms "water," "aluminum," "topaz," "quark," and "polio"? What concepts would these words express? TWD entails that either they would express no concept or they would express a concept different from the one they actually express.

I will argue that neither alternative is acceptable to the externalist. The argument concerning the first option is straightforward. There do exist plenty of empty kind terms, terms that have the basic character of kind terms (whether natural kinds or not), that do express concepts, but that lack an extension. Given this, there is nothing in the nature of a kind term that requires it to be nonempty. So there is no barrier to supposing that many nonempty kind terms, terms that should be treated as world-dependent by the externalist, might counterfactually be empty yet meaningful. As we will see, considering the relevant counterfactuals involving specific instances bears out the conclusion.

The argument concerning the second option is more complex. I will spell it out shortly. But here is the basic thought. Take a nonempty kind term. Suppose that it lacks an extension. Consider what it means, what concept it expresses. The concept it expresses must be what you get when you take the original nonempty concept and remove its extension. But this is just a non-world-involving concept, one that is expressed by the word in both the actual and counterfactual circumstance. So if there is any difference between the concepts expressed in the actual and counterfactual cases, it can be at most a difference of extension conditions. But this difference in extension conditions does not affect the world-independent content that is shared across the two cases.

Schematically, the argument proceeds as follows. Take twin planets TE_1 and TE_2. Take twin subjects or groups on the two planets. Let TE_1 subjects use the word W_1 and TE_2 subjects use the word W_2, as if they were kind words. W_1 and W_2 have the same phonographic form (pronunciation and spelling, after Kaplan 1990b) and the same use in all individualistically specifiable respects. Let W_2 apply to some real samples on TE_2. Let W_1 be empty in TE_1: either it lacks an extension, or it extends only over samples that are not present on or near TE_1. The externalist now faces a dilemma. He must either deny that W_1 expresses a concept or concede that it does. But the former option is obviously not acceptable. And the latter option is not acceptable to the externalist. For if W_1 expresses an empty concept, that concept will also be expressed by W_2, and there will be a shared world-independent content. That is the argument skeleton. Now I put flesh on the bones.

Empty concepts

It is possible that the more courageous externalist will want to choose the first of the two options offered by the argument skeleton and deny that the target kind terms could lack an extension while yet being meaningful. The most courageous (and arguably the most consistent and honest) will want to hold this on general grounds, grounds that apply to the majority of real empty kind terms. They would hold that natural-kind concepts are essentially relational. It is of the essence of our natural-kind concepts that they put us into cognitive contact with real kinds in nature. Hence empty kind terms do not express concepts at all. They are just the output of a malfunctioning concept-forming mechanism (to borrow an expression from Ruth Millikan [1984, 1989]).

This most courageous of externalist views is catastrophic. There are numerous empty kind terms that we must take to express concepts. I will dwell on this point, the importance of which is typically underestimated. Empty concepts, whether of kinds or individuals, cannot be swept under the rug. Far from being a rarity, they are a pervasive and significant feature of human cognition.

Religion and spiritual belief play very important roles in our world. All over the world, since the earliest periods of human existence of which we have knowledge, people have worshiped gods and lesser spirits, tried to appease them, blamed them for disasters, and used them to explain features of the world around them. Here is a brief summary of common religious ideas from Pascal Boyer, an anthropologist who specializes in the topic:

In all human groups one can find a set of ideas concerning nonobservable, extra-natural agencies and processes.... It is assumed in many (but not all) human groups that a nonphysical component of persons can survive after death.... It is often assumed that certain people are especially likely to receive direct inspiration or messages from extra-natural agencies such as gods or spirits.... In most (not all) human groups it is assumed that certain salient events (e.g. illness or misfortune) are symptoms of underlying causal connections between supernatural beings and the world of the living. (Boyer 1994)

Most human groups believe in ghosts: nonphysical entities endowed with various psychological characteristics and powers to affect the physical world. But ghosts do not exist. The terms "ghost" and "spirit" and the various specific terms used by particular cultures for particular kinds of ghosts are empty. Interestingly enough, though, "ghost" has the basic character of a natural-kind term. In a way, of course, it is precisely an *un*natural-kind term, in the sense that while some characteristics of ghosts are familiar intuitive ones (such as their psychological characteristics: ghosts perform modus ponens, just as we living people do, etc.), many of their properties run precisely counter to intuitive expectations (Boyer 1994). Nevertheless, the terms typically used to pick out specific kinds of ghosts by specific cultures exhibit the core features of kind terms. For example, the properties of ghosts are learned by inductive generalizations over the kind. Here is Boyer, discussing bekong, the ancestor ghosts of the Fang (a people who live mainly in Gabon and Cameroon): "Subjects spontaneously assume that all or most ghosts have the powers that are exemplified in particular anecdotes or stories. This would not be possible, without the prior assumption, that ghosts are, precisely, a kind, that one can safely produce instance-based general principles" (1994, 400). Presumably, although Boyer does not discuss this, other core features of natural-kind terms are also present. Bekong do not form an observational or artifactual or nominal kind. Rather, I presume, the Fang regard them as real entities with natures that are partly unknown but to an extent discoverable. Externalist Fang would regard the extension of the term "bekong" as determined by the facts about bekong, not by stipulation or subjective judgement. If it turned out the Fang were right, then "bekong" would also generate Putnam's Twin Earth intuitions, much as "tiger" does.

It is not a serious option to doubt that the Fang have an empty concept *bekong*. It certainly appears that the concept plays an important role in their thought. It features in their explanations of various phenomena (sickness and so on). And it features in beliefs, desires, and other cognitive states that motivate many of their activities. Moreover, Boyer spends some time discussing how Fang children acquire the concept *bekong*. Boyer begins to disentangle the contributions of innate components, inductive generalizations, explicit teaching, and so on, just as developmental psychologists do when they study the acquisition of concepts in other cases.

John McDowell, in the course of defending the thesis that certain singular concepts are world-dependent, discusses a fictional case of a native saying, "Mumbo Jumbo brings thunder." McDowell holds the courageous view that, except in special circumstances, empty singular concepts are impossible. Hence the native is not, on his view, saying anything or expressing a belief. McDowell says, "In practice, an interpreter might say things like 'This man is saying that Mumbo-Jumbo brings thunder,' and might explain an utterance which he described that way as expressing the belief that Mumbo Jumbo brings thunder. That is no real objection. Such an interpreter is simply playing along with his deluded subject—putting things his way" (1977, 175). Segal (1989b) responded that putting things the native's way is good (commonsense) psychology: put them any other way and you miss the point.

More important, however, attribution of empty concepts is not just empathic commonsense psychology. It is part of serious scientific work, such as that in which Boyer is engaged. Indeed, in Boyer's case, it is an explanandum (how do people acquire concepts of ghosts?), as well as a component of explanation. (See also Barrett and Keil 1996 for a study of concepts of God.)

As emphasized above, the Fang and their concept *bekong*, far from being an unusual case, are typical of human cultures. Think of the time and energy spent by the ancient Egyptians on their temples to the gods and their preparations for life after death in the spirit world. We can make sense of their activities if, but only if, we allow that they have genuine concepts (*god*, *spirit*) that play a major role in their cognition and action.

The same moral applies to scientific thought. There is, for example, the ether, described by the *Oxford Concise Science Dictionary* as follows: "[A] hypothetical medium once believed to be necessary to support the propagation of electromagnetic radiation. It is now regarded as unnecessary.... The existence of the aether was first called into question as a result of the Michelson-Morley experiment." If scientists had no concept of the ether, then it is hard indeed to see how they could once have *believed* that it was necessary for the propagation of electromagnetic radiation, later *called into question* its existence, and then come to *regard* it as unnecessary.

And, to cite another standard example, we have "terra pinguis" or "phlogiston," which was supposed to explain combustion. Here is the *Oxford Concise Science Dictionary* again:

The existence of this hypothetical substance was proposed in 1669 by Johann Becher.... For example, according to Becher, the conversion of wood to ashes by burning was explained on the assumption that the original wood consisted of ash and *terra pinguis*, which was released during burning. In the 18th century George Stahl renamed the substance *phlogiston* ... and extended the theory to include the calcination (and corrosion) of metals. Thus metals were thought to be composed of *calx* (a powdery residue) and phlogiston.

The theory was finally overthrown by Antoine Lavoisier in the late eighteenth century. Again, it would be difficult to explain the evolution of the theory of combustion without attributing a concept of phlogiston to the scientists involved.

A case that seems to me somewhat intermediate between science and religion is commonsense psychology. It is a powerful explanatory theory, largely correct, better than any scientific psychology or neurology, in many areas of its explanatory and predictive domain. Or so I would argue (following Fodor 1987, chap. 1). But it strikes me as plausible that in its basic form it is profoundly dualist, treating minds as immaterial spirits, the sorts of things that can transmigrate, exist disembodied in the spirit world, and so on. (A similar view is argued in detail in Wellman and Estes 1986, and Wellman 1990). Not only do many people actually believe in these possibilities, but also every normal human can entertain them with the utmost ease. Indeed, we retain this capacity even after we have become materialists and believe that psychological properties are properties of material things. This is a striking fact. For all that, immaterial spirits, souls, do not exist. Our concept of immaterial minds has no extension.

The main argument for attributing empty concepts in all these cases (ghosts, ether, etc.) is simply that by so doing, and only by so doing, can we make psychological sense of a very wide variety of human activity and cognition. Anthropologists, historians of culture and science, psychoanalysts, and others actually do this. In so doing, they provide what appear to be perfectly cogent psychological explanations. If these explanations have apparent flaws, these flaws are not traceable to the attribution of empty concepts. So well entrenched and successful is this practice that it would require a powerful argument indeed to show that it was deeply problematic or incoherent.

Moreover, not only do we standardly offer explanations involving attributions of empty kind concepts. We have no viable alternative. One cannot simply dispense with such attributions and be left with any workable intentional explanations of the various phenomena. Nor can any nonintentional theory address the phenomena. Appeal to neurology or purely syntactic computational theory would be an expression of pure faith. We have no idea how such research programs could bear significantly on the psychological, cultural, anthropological, and historical phenomena that we seek to describe and explain. Nor have we any reason to suppose that one day they will.

Let us allow, then, that humans have or have had concepts of phlogiston, ether, immaterial souls, etc. This conclusion, which appears mandatory, already invites a strong argument against TWD. For if these terms express concepts, why shouldn't it make perfect sense to suppose that "quark," "helium," "Gila monster," "polio," or even "water" might, counterfactually, be empty yet express a concept?

The only barrier to the coherence of these counterfactuals comes from the environmental conditions involved. For example, we cannot seriously imagine people in a world with no water: as noted above, we are largely made of water. Moreover, even if we ignore that fact, it is hard to spell out the case without running into excessive contrivance (mad scientists with mind-bogglingly advanced apparatuses and the like). The contrivance makes it difficult to assess the salient points of the counterfactual, such as what the extension conditions of the protagonists' concepts would be.

Nevertheless, such counterfactuals are compatible with the nature of the concepts. For there is nothing in the nature of kind concepts themselves that requires there to be an actual extension. Given this fact, how can we resist the conclusion that if, so to speak, one were to remove the extension from a concept, one would be left with some real content, a content shared across the empty and non-empty cases? Or, to put it less metaphorically, how can we resist the conclusion that each of these words expresses a content that could exist in the absence of its extension? I will argue that we cannot.

From the empty to the full

The next phase of the argument is best developed in the context of a specific example. Myalgic encephalomyelitis (ME, also known as "chronic fatigue syndrome" or "CFS") is a condition characterized by chronic tendency to extreme fatigue. Symptoms include dizziness, rashes, aches, sensitivity to light and sound, cold sores, swellings, forgetfulness and many others. At the time of writing, there is disagreement in the medical profession about whether ME is caused by a virus and, if it is, how the process occurs.

Let us suppose that it is epistemically possible that there is no such thing as ME. The following is then an epistemically possible future. It turns out that there are a wide variety of different causes of the symptomatology associated with ME: viruses, stress, hysteria, dental amalgam, educational practice (all of which have been seriously suggested). The medical establishment stops using the term "ME," since it is unhelpful for taxonomic purposes. Not only is there no common underlying ailment, but there is also no reason to count ME as worthy of recognition as a syndrome. Doctors become disposed to say: "Some people used to talk about a condition called 'ME,' but it turned out that there wasn't really any such thing."

But it is also epistemically possible that "ME" will become a natural-kind term. Doctors might discover that there is a single virus responsible for most of the diagnosed cases, that most of the diagnosed patients that turned out not to suffer from the viral infection had a different symptomatology, and so on. "ME" could come to refer to the disease caused by the virus, and to nothing else.

In both cases doctors and patients use the term "ME" to engage in apparently cogent conversations. For example, a typical subject, Peter, finds that he is becoming more and prone to fatigue. He becomes worried and goes to his doctor. After consultation, the doctor says, "You have ME," and goes on to explain the current wisdom. Psychological explanation of both doctor and patient requires attributing the concept of ME. We might say, "Peter took some time off work because he believed that he suffered from ME and he desired to alleviate the symptoms of ME and he believed that by taking time off work," etc. And we can explain why the doctor thought that Peter had ME and why she went on to prescribe action that she thought might alleviate the symptoms of ME.

Let TE_1 be the empty case and TE_2 the nonempty one. Let our twin subjects be $Peter_1$ and $Peter_2$. We must allow that $Peter_1$, in TE_1, expresses a genuine concept by his term "ME." If we do not, then we will have no adequate explanation of his words and deeds. Call this concept "C_1." Let C_2 be $Peter_2$'s concept and let "ME_2" be our word for expressing C_2. So "ME_2" means just what "ME" actually means if our world happens to be TE_2. And let C_1 and C_2 be individuated by their cognitive content. I will now give three arguments that $C_1 = C_2$, that is, that Peter$_2$ has the same "ME" concept, a concept with the same cognitive content, as does Peter$_1$.

First argument

The first argument concerns the cognitive roles of C_1 and C_2. The cognitive role of a concept is its causal role, as specified by a psychological theory, its causal role relative to other psychological states and actions. Here are some aspects of the cognitive role of C_2: Peter$_2$ came to believe that he was suffering from ME$_2$, because his doctor said, "I fear you have ME," and Peter$_2$ believed his doctor. Peter$_2$'s belief that he has ME$_2$, along with his belief that sufferers from ME$_2$ often need to take time off from work, caused him to fear that he would need to take some time off from work. Peter$_2$'s desire to be cured of ME$_2$, or at least to have some of its symptoms alleviated, caused him to seek further advice from his doctor.

Now suppose that a psychologist studying both Peters simply assumes that $C_2 = C_1$. That is, she assumes that whatever concept it is that Peter$_1$ has, with its particular cognitive content, is the very concept that Peter$_2$ has and expresses with his term "ME." The assumption would work perfectly well because C_1 is just the sort of concept that can play the required cognitive role. We know this because that is exactly the role it does play in Peter$_1$'s psychological economy. If, as the psychologist hypothesizes, Peter$_2$ did indeed possess the concept C_1, then he would reason and act exactly as he does. Attributing the concept C_1 to Peter$_2$ would therefore work perfectly well in psychological explanations. And this provides a reason for thinking that the attribution would be correct.

Of course, the subjects' exercises of their concepts do have different effects in their different environments. For example, $Peter_1$ consults one doctor on TE_1, and $Peter_2$ consults a numerically distinct doctor on TE_2. But it is natural to explain this in terms of the differences between the twins' locations rather than in terms of differences between their concepts. For if $Peter_1$ were to take $Peter_2$'s place on TE_2, he would interact with the environment exactly as $Peter_2$ would have if $Peter_2$'s place hadn't been usurped (see Fodor 1987, chap. 2).

Notice that the argument is not intended to show that there is no use for wide content. There might be generalizations one could frame in terms of the concepts' wide contents that would not be captured in terms of the cognitive content shared by both twins. It might be, for example, that $Peter_2$'s concept shares a wide content with the concepts of other inhabitants of TE_2 but not with C_1. And these shared wide contents might enter into generalizations concerning the relations between subjects and the variety of ME that is present on TE_2. These generalizations would not be captured by attributing C_1, individuated by its cognitive content, to $Peter_2$. I believe that any such explanatory shortfall could be made up by supplementing explanations in terms of C_1 with specifications of $Peter_2$'s relations to his environment. But the point is not obvious and would require further argumentation. For although there would always be some explanation of why someone with C_1, appropriately situated in TE_2, would interact with TE_2 just as $Peter_2$ actually does, the explanation might be pitched at a relatively unilluminating level of description. It might be that the duplex explanations framed in terms of C_1 and environmental relations would be less satisfactory than explanations that could avail themselves of wide content.[7]

Rather, the point of the argument is that standard psychological explanations, such as belief-desire explanations of action or of the formation of further beliefs and goals, could be successfully formulated in terms of a cognitive content that is common across the twins. And the hypothesis has an obvious virtue. For it offers a compelling explanation of the twins' evident psychological similarities. The reason that the twins' "ME" concepts have the same cognitive roles would be that they have the same cognitive content. If the two concepts had only wide contents, then it is not at all obvious why their cognitive roles should be the same. TWD would therefore have a lot of explaining to do.

The argument is not conclusive, since the success and virtue of an explanation do not guarantee its truth. The explanations are successful because $Peter_2$ thinks and acts just as he would if he had C_1. It does not follow from this that he actually does have C_1. But I think it follows that we have good reason to think that he does: there would be considerable evidence in favor the hypothesis. Further substantiation of the claim would depend on an account of the shared cognitive content. If we could formulate a good account of a kind of cognitive content that would be shared across twins and show how such content could feature successfully in psychological explanation, then the claim would be largely vindicated. If, by contrast, attempts to formulate the account were perpetually to run into difficulty, then we would have reason to think that the hypothesis was on the wrong track. Such matters will occupy chapters 4 and 5.

Second argument

For this argument, I need to assume that concepts have supervenience bases. For simplicity I will assume that these are physical, by which I mean nonmental, in some loose sense. The claim is that if a subject S has a concept C, there exists some set of physical properties of S, and possibly of relations between S and her environment, that are necessary in the circumstances and sufficient for her possession of C. To put it more roughly (and only roughly equivalently), there is some minimal set of physical properties and relations in virtue of which S has C. If you duplicated these properties and relations, you would get a counterpart of S who would also have C. I will call this set of properties and relations, "the base" (short for "the supervenience base of C"). The base might include such things as particular configurations of synapses, higher-order functional properties of these configurations, and any external facts you like concerning the evolution of species, causal relations between S and her environment, and so on.

The assumption of a base might exclude some philosophers from the discussion. But I envisage arguing with an externalist who is happy to agree that content supervenes on something. The dispute is whether that something is internal to the subject or involves external matters too.

I proceed with the argument. Consider TE_1, the empty case. It is allowed that $Peter_1$ has some concept C_1, expressed by his word "ME." $Peter_1$'s possession of C_1 has a base: there exists some set of physical properties and relations in virtue of which $Peter_1$ has C_1. Call this set "$Base_1$." If you duplicate $Base_1$, then you get a counterpart of $Peter_1$ who also has C_1. But $Base_1$ is duplicated in TE_2. $Peter_2$ is a counterpart of $Peter_1$. And whatever physical properties and relations suffice for $Peter_1$ to have C_1 apply also to $Peter_2$. So $Peter_2$ has C_1. But if he has C_1, then C_1 is expressed by his uses of "ME." So $C_1 = C_2$. So the two Peters' "ME" concepts differ at most in their extension conditions.

Let us go through this argument more slowly. The main claim is that $Base_1$ includes nothing relevant that is not present in TE_2. I have no conclusive argument for this. But it does seem reasonable. The only difference between TE_1 and TE_2 is that where TE_2 has a virus, TE_1 has nothing much. Where in TE_2 the symptoms of ME are caused by a specific virus, in TE_1 they are caused by a motley of different phenomena. Now, it is not this motley of different phenomena that account for $Peter_1$'s possession of the concept. By assumption, his concept is empty. It has no extension. It does not apply to the motley of phenomena. But since the content of C_1 does not depend even in part on local instantiations of its extension, what with there not being any, it must depend on something else. Well, what would that be? It could be something about $Peter_1$'s neural states, the functional role of his Mentalese word, his relations to the symptoms associated with ME, his relations to his doctor. But all of these things are duplicated on TE_2. Everything present on TE_1 that could be relevant is also present on TE_2. For where TE_2 has a presence, the virus, TE_1 has an absence. There is nothing of note present on TE_1 that is absent on TE_2.

So $Peter_2$ also has C_1. It is obvious that C_1 is expressed by $Peter_2$'s uses of "ME." It surely is not expressed by any other word of his. It must be instantiated in the same neural states in both Peters. It must have parallel causal roles in the two.

So the externalist is forced into a two-factor theory. When $Peter_2$ says "ME," he expresses C_1, the very same concept that $Peter_1$ expresses. If there is any room left for a distinction, it can only be that when $Peter_2$ says "ME," the concept he expresses, C_2, has some content additional to that of C_1 considered alone: its extension or extension conditions. But that is incompatible with TWD.

Third argument

The third argument is similar to the second but exploits a different sort of condition sufficient for concept possession. The base constitutes synchronic conditions. What is it about $Peter_1$ now in virtue of which he has C_1? Answer: $Base_1$. But we can also consider diachronic, developmental conditions. Specifying these would answer the question, How is the concept ME acquired? The answer would describe the innate endowment and environmental conditions that explain acquisition of the concept.

The argument runs more or less as before. Consider $Peter_1$. How did he acquire his "ME" concept? He learned about ME from the doctor. But that is only the last part of a long story. When he encountered the doctor, he was already in a position rapidly to acquire the concept simply by hearing a few words. In order to explain this, we would have to develop a complex theory of his innate endowment and his developmental history: how he learned about diseases and so on. But the key point is, whatever the truth of the matter, everything specified by that theory would be present on TE_2. There is nothing available on TE_1 to explain how $Peter_1$ acquired C_1, his empty concept of ME, that is not also present on TE_2.

It follows that a theory specifying conditions sufficient for the acquisition of C_1 is applicable to $Peter_2$. But then $Peter_2$ has C_1 as well. And C_1 is the concept he expresses by his word "ME." So the difference between $Peter_1$'s C_1 and $Peter_2$'s C_2 concerns at most the extension.

The externalist might object as follows. The complete theory of how $Peter_1$ acquired his "ME" concept would specify a condition that obtains in TE_1 and not TE_2. Specifically, it would mention the absence of any disease or syndrome lying at the ends of the causal chains that lead to uses of "ME." The acquisition conditions include many factors that are common across TE_1 and TE_2. But those common conditions act only in the context of their specific environments to endow their subjects with concepts. When these conditions obtain in TE_1, they result in $Peter_1$'s acquiring his "ME" concept, and when those same conditions obtain in TE_2, they result in $Peter_2$'s acquiring a different "ME" concept.

The objection is coherent, but implausible. It is also out of line with psychological practice. Developmental psychologists do not usually care whether the concepts they study are empty or not. If they believe that a concept is empty, they do not typically mention this fact. I doubt, for example, that Boyer ever explicitly points out that ghosts do not exist. And Henry Wellman, studying the acquisition of mentalistic concepts (concepts of belief, desire, perception, and so on) specifically states that he abstains on the question of whether these folk-psychological concepts are empty (1990, 151–153).

The bekong

There are other kinds of examples that it might be fruitful to develop. In the ME case, I left it open whether the term actually applies to a real disease in the real world. We might also consider cases with an expression that clearly does have an extension exemplified in the real world, such as "electron" or "polio." The counterfactual situation would then be one in which the term is empty, but as much as possible is held fixed. Other cases would reverse the procedure. We would take an expression that clearly lacks an extension in the actual world, such as "ghost," "unicorn," or "phlogiston." We would then imagine a counterfactual situation in which the term had an extension. Counterfactuals of both kinds can be difficult to handle, since they tend to require breaches of laws or a very large stretch of the imagination. An advantage of the ME case is that it avoids these difficulties.

I will go through just one more example that runs more or less as smoothly as the previous one, even though it involves a slightly greater exercise of the imagination. After that, I will consider the extent to which the argument schema generalizes to other kinds of cases and assess the ramifications for the larger debate between internalism and externalism.

Since we have already been introduced to the Fang and their concept *bekong*, this is an appropriate choice for our next case.

The bekong are spirits who dwell in invisible or hidden villages and breed wild animals. Strange events are often interpreted as signs of bekong activity: a fleeting shadow in a clearing, a chased animal suddenly disappearing in a bush. There is considerable ambiguity concerning the exact process that transforms people into bekong. Many believe that it is the person's *nissim* (shadow) that leaves the body and becomes a bekong. ("Nissim" is often used as a metaphor for individual identity.) Bekong are invisible and intangible. Encounters, such as fleeting shadows or disappearing animals, are interpreted as a ghost's wanting to send a signal to the living or to be noticed. But serious encounters or conversations with bekong occur only in dreams or trances. Bekong move fast and can pass through physical objects. But they cannot be in two places at once. (All of this is from Boyer 1994.)

In our world, we can assume that the Fang are mistaken. There are no bekong. But we can imagine a Twin Earth in which the descriptions associated with bekong do apply to real entities. I take this to be a conceptual possibility. I think it is more or less nomologically possible, although there are anomalies that would need explanation. In particular, as Boyer mentions, it is hard to see how these apparently noncorporeal entities can interact with the physical world, how they can eat and drink (as they are supposed to), and so on. Perhaps we should suppose that on this twin Earth, the term "nonphysical" would not be appropriate for bekong. Maybe we can suppose that they are composed of a special kind of energy, unknown on Earth, that nevertheless interacts with other forms of energy in systematic, quantifiable ways.

Twin Fang, then, have a kind term "bekong" that applies to real entities. It is a bona fide natural-kind term. One way to see this is to imagine two twin Earths in which "bekong" applies to real individuals but the underlying nature of the bekong on the two planets is very different. The nature of the energy is different, governed by different laws. Moreover, the mysterious process by which humans, at death, are transformed into bekong is also different. In one case the bekong has been present all along, the main component of the living human's psyche. In the other, the bekong only appears at death, via a complicated process of transformation.

This case, involving the two twin Earths with different kinds of bekong, is parallel to the original Twin Earth case. If one favors Putnam's account of "water," then one ought to favor the same account for these twin bekong. The Fang are like Earthlings of 1750 in respect of water: competent lay users of the relevant term, but not scientifically expert. If Putnam's account is right, then when scientists discover the truth about bekong on one planet, the term will probably come to denote only the entities on that planet. Their term "bekong" would not be applied to the counterparts on the other planet. Counterparts are not instances of the "same ghost" relation, where the nature of this relation is to be defined by science.

I predict that many readers (or at least many nonphilosophers) would not have Putnam's intuitions about this case. I predict that the leading intuition is that the Fang of all three planets have the very same concept *bekong*. This is all to the good, since the case is in fact parallel to Putnam's original. Hence it only goes to bring out the unreliability of externalist intuitions. But now I want to turn to the argument for sameness of content, rather than the intuition.

The argument runs exactly as it did for ME. Take the actual, Earth Fang concept *bekong*. Twin Fang too must have this concept. The reasons are as before. First, psychological explanations given of the actual Fang (such as those given by Boyer) transfer to twin Fang without anything going wrong. Everything Boyer says about the Fang could be said about twin Fang, and explanation and prediction would proceed without hitch. Further, whatever accounts for real Fangs' possessing the concept is present on Twin Earth too. This applies to the theory of concept acquisition, in terms of innate endowment, inference, and so on. And it applies in terms of the base, the synchronic conditions sufficient for possession of the concept. Thus twin Fang meet two sets of conditions sufficient for possession of the same concept as real Fang. The conclusion is that if there's a difference between the real empty concept *bekong* and the counterfactual nonempty concept, it concerns at most the extension conditions.

Harder cases

The two examples developed above, with ME and bekong, run smoothly in part because, in the nonempty case, instances of these two kinds are not easily observable by the protagonists. More specifically, apparent perceptions of instances of bekong or ME play little or no role in acquisition of the concepts. Any example sharing this feature seems to proceed similarly. Thus one might choose quarks, phlogiston, or gravitational fields and come up with similar results. Equally well, one could choose an example involving a kind whose instances are observable but happen never to have been observed by the protagonists: the concept of diamonds or aluminum or fish possessed by a subject or group who have never actually encountered a sample but have only seen pictures or heard tales.

Matters proceed differently when protagonists do regularly perceive instances. When we come to the part of the story that focuses on the empty concept, we are forced to engage in some artifice. For we have to describe suitable empty counterfactual counterparts of occasions when protagonists perceive instances of the kind in the nonempty case. But this is likely to require radical departures from the nonempty case: we must consider brains in vats; subjects of experiments run by technologically advanced brain-manipulating scientists; so called "swamp people" (after Davidson 1987), duplicates of humans who arise, fully formed, as the result of an extraordinary quantum accident; or some other such bizarrerie. In such cases, the crucial step of the argument, claiming that the subjects in the empty case possess a concept, becomes more difficult to motivate.

I think the main reason for this is simply that the condition of the subjects is too far from what we are familiar with. It is not only that we lack intuitions about the cases. Or if we don't lack intuitions, I think we should. But more important, the situations are so different from mundane ones that it is not so easy for us to know to what extent a psychological theory could get a grip. The main difference concerns concept acquisition. When I argued in general that we must allow the existence of empty concepts (*bekong, god, phlogiston*, and the rest), I was talking about real people with real histories of acquisition. But brains in vats and swamp people have nothing analogous:

the brains are just bulldozed into the relevant states either by accidents or by scientists with electrodes. Because of this, the dialectic is more complex. I will consider a difficult case and explain how it relates to the general argument.

Earth, 1750. A tribe of Bedouins ride camels in the desert. They have no water. Nor can they see any. But there are some dunes ahead, and they believe that they will find water beyond the dunes. "There is water beyond the dunes," they say in Arabic. Twin Earth this time is very different from Earth. On it there is neither water nor any kind of Twin water, nor has there ever been any. It is Dry Earth (the term is from Boghossian 1997). But there is a stretch of desert exactly resembling that in which we found the Bedouins. Due to an extraordinary quantum accident, Twin Bedouins have suddenly appeared there. (We can either forget that Bedouins are made of water or suppose that the Twin Bedouins are made of water too.)

The Twin Bedouins utter the word forms, "There is water beyond the dunes" (or the Twin Arabic equivalent). And they head rapidly in the expected direction. To run my argument again, I would need to claim that these Twin Bedouins express a concept with their word "water." Of course, I do claim this. But the claim is less plausible than its analogue for ME and bekong. The reason is, as suggested above, that these Bedouins have nothing resembling a history of acquisition of the concept. The externalist might now simply reject my claim and try to block my argument with respect to this particular case. There are number of reasons why this would be a bad move.

First, the main argument from before remains in place. If we simply assume that the twin Bedouins are subject to psychological explanations, we can provide a good account of their behavior. This account would have all, or at least most, of the virtues of psychological explanation generally. Insofar as the attribution of a concept in the explanation of action gains explanatory power by linking such attributions to the history of the subject's acquiring the concept, such explanatory power is absent in this case. But a great deal of the power of psychology does not depend on this relation to acquisition. If, for example, we supposed that the twin Bedouins were expressing a belief by their words "There is water beyond the dunes," we could predict that they would head toward the dunes. All such forward-looking explanations would remain workable. The predictive power of psychology would remain intact. Hence the conclusion from before remains in place: if we attributed concepts to the Bedouins, our so doing would achieve explanatory and predictive power. Hence the attributions would earn their keep and prove themselves.

Second, it would be unprincipled for the externalist simply to deny at this point that these twin Bedouins had some concept expressed by their term "water." If what I have argued hitherto is correct, the externalist has no grounds on which to make this claim. There is no barrier to empty concepts in general. Therefore, interaction with instances of the kind is not in general a necessary condition for concept acquisition. So why would the twin Bedouins lack any "water" concept? I take it that the externalist will allow that one could acquire the concept of water from books, pictures, other people, or, if one were a brilliant theorist, one might come up with a complex theory that could endow one with the concept. I take it that considerations already discussed indicate that one can acquire an empty concept via inference, interaction with others, and so on. But then why is concept acquisition restricted in principle to these kinds of processes? What unites these candidate conditions of acquisition but excludes the conditions of the twin Bedouins?

What the externalist needs at this point is a principled argument for the view that acquisition of a concept requires some specified kinds of interaction with the environment, all of which are absent in this case. But given the heterogeneous ways in which one can acquire a concept, this might prove difficult. Thus the dialectically weak position in which the externalist seems to be is this: she cannot argue that any particular disjunct of a disjunction of heterogeneous ways in which one might acquire a concept is necessary for acquisition, but the disjunction as a whole is necessary. That seems unprincipled.

Once it is allowed that the Bedouins express some concept or other when they say "water," most of the arguments go through as before, with the conclusion that the real Bedouins express that concept too. The only difference is that the argument from acquisition no longer applies.

Externalism and empty concepts

Before I conclude the main argument against TWD, it is worth briefly raising a final difficulty for the thesis. The difficulty is that it is not easy for the externalist who allows for empty concepts such as *bekong* or *phlogiston* to give a plausible account of their extension conditions. What, according to the externalist, are the extension conditions of these concepts? Only two responses suggest themselves. The first is that such concepts are modally empty: they don't apply to anything in any possible world. The other is that they are motley concepts, applying, roughly speaking, to anything satisfying the core descriptions associated with the them. Let us consider these in more detail.

The first option makes lack of extension essential to a concept just as externalists think that extension is essential to nonempty concepts. The idea is this. The externalist claims that Oscar's "water" concept is true of water (H_2O) not just in the actual world but across all possible worlds. That is why it is true to say that Oscar's concept would be false of a sample of XYZ, if there were one. Extending this line of thought to empty concepts would entail that the twin Bedouins' "water" concept also fails to apply to XYZ, or to H_2O, or to any counterpart of water. Its semantic relation to nothingness mirrors the semantic relation of nonempty concepts to their real extensions. If a twin Bedouin were to say, "There is no water, but there might have been," he would be saying something false in his language. No matter what kinds are possible relative to Dry Earth, none of them would be in the extension of the twin Bedouin word "water." (Compare Kripke [1980, 24] on unicorns.) There might be phlogistonoid substances and ghostoid entities in various possible worlds, but these are not phlogiston or ghosts, any more than XYZ is water.

The problem is that this result is plainly counterintuitive. Although there are no ghosts, there might have been. So arguments for externalism that are based on intuitions about Twin Earth are considerably weakened. For now the intuitions are going the other way.

The second option is the one that accords with intuition. This would be the idea that the word "ghost" would apply to any suitably ghostly entities. There are thus many different possible kinds of ghosts in different possible worlds. The trouble is that it would be ad hoc for an externalist to allow this. It is the first option that conservatively extends the normal externalist line on nonempty concepts to empty ones. The motley concept, the one that applies to any suitably ghostly entity, is the one available in worlds where a motley of different kinds of ghosts lies at the ends of the causal chains leading to uses of "ghost." It is not clear that a spiritually consistent externalist should allow that the motley concept is also available in our world, where nothing lies at the ends of the causal chains.

Thus the challenge for the externalist is either to motivate rejecting the intuitions against option one or to motivate the departure from spiritual consistency required by option two.[8]

Conclusion

The main argument thus far has shown, I believe, that the thesis that natural-kind concepts are world-dependent is false. For we can take any concept C that is alleged to be world-dependent and run the argument to show that C has the cognitive content that it would have had if it lacked any extension. If it can be done for *water*, it can be done for *aluminum*, *tiger*, *topaz*, and all the rest.

2.3 WEAKER EXTERNALIST THESES

The argument against TWD does not bear directly on social externalism, which must be dealt with separately. However, I think it disposes of the most popular and influential form of nonsocial externalism about kind concepts. True, one can think of various forms of such externalist theses that are not vulnerable to the argument as stated. There is the thesis, for example, that while, say, a *water*like concept need not have any extension, such a concept would only be available in a world with either sand or air and thus the concept of water is indirectly world-dependent. But this would be ludicrous.

My argument does not directly address forms of externalism that do not require concepts to stand in real relations to samples of their extensions but that do see more general relations to the environment as essential.[9] I cannot here directly address all theories of this sort. But there is one milder thesis of world dependence that can be dealt with relatively swiftly.

Fodor's externalism

Suppose that one held (following Fodor 1994) the following sort of causal theory of content. Assume there is a language of thought (LOT). Let C be a syntactic item in LOT. Then, roughly speaking, C means K if it is a law that any K would cause a token of C in the right circumstances. This means, roughly, that if a K were to occur in the right circumstances (good light, etc.) the subject would think a thought of the form *Oh, there seems to be a C there.* Assume that there could be a noncircular account of the right circumstances. And be aware that Fodor has other important clauses that I have left out.

Now, Fodor claims that this is a somewhat externalist theory. The reason (the one that matters for present purposes) is that the relevant counterfactuals of the form "It is a law that K would cause C to token" are themselves world-dependent. For to evaluate a counterfactual relative to a given world, one considers what happens in worlds close to it. But then, if we take two twin Earths, TE_1 and TE_2, it may be that they are in different modal neighborhoods: worlds close to TE_1 may be far from TE_2 and vice versa.

Fodor considers the specific case of a swamp man on XYZ Twin Earth. He says "If [the swamp man's] words and thoughts are about XYZ," which Fodor thinks they are, "that's *not* because he's causally connected to the stuff" because, recall, he has only just come into existence, so no causal connections have yet come into play. "Rather it's because *it's XYZ that would cause his 'water' tokens in all the worlds that are nearest to Twin Earth*, there being ... no H_2O on any of them" (my emphasis).

So, since XYZ worlds and H_2O worlds are so far apart, they are in different modal neighborhoods, and the relevant counterfactuals are different. That seems to be the idea. If so, it is a mistake. The relevant counterfactual is not "What would cause 'water' to token in nearby worlds?" If that were the case, we could never talk about distant possible worlds at all. For the content-determining counterfactuals would only concern nearby worlds and never stretch far away. The point about nearby worlds is quite different. The counterfactual relevant to content is "What would cause 'water' to token?" If we want to know whether H_2O would cause this swamp man's 'water' to token, we consider not the possible worlds near to Twin Earth, but the possible worlds nearest to Twin Earth on which H_2O and swamp man are together in the right circumstances. If H_2O worlds in which H_2O causes his 'water' to token (in the right circumstances) are closer to Twin Earth than H_2O worlds where H_2O does not cause his 'water' to token (in the right circumstances), then H_2O is in the extension of his 'water' after all. And, of course, they are and it is.

The result generalizes to the cases featured in my argument. Consider, for example, ME and TE_1 and TE_2. Is the virus on TE_2 in the extension of $Peter_1$'s "ME"? Fodor's theory suggests that it is. For on the nearest worlds to TE_1 on which the virus occurs, it would cause a token of his concept C. At least it would "in the right circumstances." Equally, if we consider any TE_3 on which there is a different virus causing the symptoms of ME, it will be in the extension of both $Peter_1$'s and $Peter_2$'s "ME."

Indeed, it would surprizing if the result failed to generalize to all twins and all concepts. Since twins are twins, their concepts will token in just the same circumstances in just the same possible worlds. Counterfactuals true of one twin will tend to be true of another. It might indeed be the case that twins' home worlds occupy different modal neighborhoods in the abstract, but it is hard to see how that could make a difference to the counterfactuals in question. Consider, for example, XYZ and H_2O worlds. Some counterfactuals do come out differently relative to the two. For example, "If there were no H_2O, then things would be different" is true of Earth, false of Twin Earth. But content-determining counterfactuals are of a different sort. These are of the form "If subject S were in circumstances X, then K would (would not) cause a token of S's LOT concept C." Presumably X, the right circumstances, would be constant across the cases. They should therefore screen out any possibility of differences in the truth values of the counterfactuals following from the differences between the twins' home worlds.

In chapter one we considered the externalist view that the content of natural kind concepts depends in part on the nature of the kinds themselves. We turn now to a variety of social externalism, which holds that the content of many kinds of concepts, not merely those of natural kinds, depends in part on the social environment, on the way other people use words.

3.1   CONSUMERISM

The examples

Putnam's "elm" and "beech"

We begin, once again, with an example of Putnam's (1975a). Putnam tells us he cannot distinguish elms from beeches. He cannot tell them apart by sight, nor does he know anything that is true of one but not the other—barring, of course, their names. But when he says "elm," his word is true of elms and not beeches. When he says "beech" his word applies only to beeches, not to elms. Twin Earth is again very like Earth. The difference now is that the Twin word "elm" is used to talk about beeches, while the word "beech" is used to talk about elms. When

Twin Putnam says "elm," he is speaking Twin English, so he refers to elms not beeches. Hence Putnam and Twin Putnam mean different things by "elm." Thus argued Putnam. Again, this conclusion seems naturally to extend to the concepts expressed by the words. So we have an apparent counterexample to internalism.

Why is it that Putnam refers to elms and elms only when he says "elm"? Because he intends his usage of the word to conform with that of the experts, he means to refer to those trees called "elms" by those who know what they are talking about. So an important determinant in fixing the extension of Putnam's "elm" concept is the linguistic practice of experts. On Twin Earth, the experts use the words differently, and so Twin Putnam's "elm" concept gets hooked up to beeches, not elms. Cut the pie any way you like, meanings just ain't in the head. Thus argued Putnam.

In Kaplan's terms, Putnam is a 'consumer' of language, rather than a 'creator', when it comes to "elm." The word "elm" that Putnam uses comes "prepackaged with a semantic value" that has been created by others.[1] It is this prepackaged semantic value that provides what Putnam himself means by "elm" and the concept he expresses by it. So let us call this brand of social externalism "consumerism."

Burge's "arthritis"

Tyler Burge introduced a different kind of example in support of consumerism. These cases also involve a subject who has only partial information about the meaning of the term involved. But these examples draw on the possibility of the subject believing an analytic falsehood.

A subject, Alf, appears to have a number of mundane beliefs about arthritis. It seems that he believes that he has had arthritis for years, that his arthritis in his wrists and fingers is more painful than his arthritis in his ankles, that there are various kinds of arthritis, and so on. It also seems that he believes, falsely, that he has developed arthritis in his thigh. He goes to the doctor, who tells him that, by definition, arthritis is an inflammation of the joints, and that therefore he cannot have it in his thigh. Alf accepts the doctor's informed opinion and goes on to ask what is wrong with his thigh.

Burge then tells a counterfactual story, which I will rephrase as a tale of Twin Earth. This time the key difference between Earth and Twin Earth is that on the latter planet, the medical profession uses the term "arthritis" more generally than it is used on Earth. It applies not only to inflammations of the joints, but also to certain related conditions, including, as it happens, what Alf and Twin Alf have in their thighs.

Twin Alf, according to Burge, lacks some, probably all, of the propositional attitudes that Alf has about arthritis. Twin Alf has no word which means *arthritis*; his word "arthritis" does not mean *arthritis*, but applies to the more general cluster of ailments. But Alf has the concept of arthritis, even though, for a period of one morning, he entertains an analytic falsehood: that he has arthritis in his thigh. Alf has the concept of arthritis because, like Putnam with "elm," he has some competence with the term and he defers to the relevant experts. When he comes to learn the truth, he might have thought to himself: *Ah, I thought I had arthritis in my thigh, but I was wrong*. Thus, when he knows the facts, he regards himself as having previously deployed the same concept he has now, the concept that applies only to inflammations of the joints.

Part of the interest of this example lies in the fact that "arthritis" is not a natural kind term. If Burge is correct, then the example shows that externalism has a broad range of application that extends far beyond natural kind concepts. Given this, it is very important not to treat the case as if "arthritis" were a natural kind term, as does sometimes happen in the literature.

"Arthritis" just means *inflammation of the joints*, full stop. It comes from the Greek, "arthron" meaning *joint* and "itis," connoting inflammatory conditions. Arthritis can be caused by over 200 different conditions of widely different sorts. The causes include ordinary wear and tear, injury, auto-immune problems, sickle cell anemia, lupus, gout, syphilis, tuberculosis and ankylosis. Since "arthritis" is not a natural kind term, it is important not to let one's intuitions be guided by experience of thinking about examples like "water" or "topaz."

Burge's own discussion of the case does not always fit very well with the true meaning of "arthritis." At one point, he calls it a "family of diseases." This is slightly misleading. Rather, it is a symptom of any of a large selection of diseases that do not form a family in any other sense. Moreover, Burge's description of Alf's state of mind imputes a similar misconception to him (Burge 1979, 95):

He will be relieved when he is told that one cannot have arthritis in the thigh. His relief is bound up with a network of assumptions that he makes about his arthritis: that it is a kind of disease.

It is of course possible that this is how Alf would react. But if so, then Alf, even after discussion with his doctor, would be under a misapprehension about "arthritis." It is only a kind of disease in the very loose sense in which itchy spots or grazed knees are kinds of diseases. There is no special reason why Alf should be relieved to learn that it is not arthritis in his thigh. After all, the doctor knew without investigation that Alf's arthritis had not spread to his thigh, since as matter of definition, one cannot have arthritis outside the joints. But matters of definition are unlikely to herald good news about a person's health. Alf's worry should be whether the condition causing his arthritis had deteriorated. And, indeed, it might have. Whatever was causing the arthritis might now be causing the pain in his thigh. (My investigations have revealed that Alf suffered from rheumatoid arthritis in the hip which had deteriorated, causing a referred pain in the thigh.)

3.2 AGAINST CONSUMERISM

Once one has the definition of "arthritis" firmly in mind, it becomes difficult to make sense of Alf's initial state of mind. Since "arthritis" just means "inflammation of the joints," it is hard to understand how anyone could believe that he has arthritis in the thigh. Burge would probably point out that the apparent conceptual difficulty here should be resolved once one realizes that Alf does not fully grasp what he believes. The internalist might reply that the notion of believing a thought one does not fully grasp is itself not easy to fathom. At any rate, the semantic proximity of "arthritis" to "inflammation of the joints" makes it impossible for the externalist to give an acceptable account of Alf's state of mind. The following argument exploits this difficulty.

Alf does not seem to believe that he has an inflammation of the joints in his thigh, knowing, as he does, that his thigh is not a joint. Indeed, he positively believes that he does not have an inflammation of the joints in his thigh. Or at least he would if he thought about it. But he does not positively believe that he does not have arthritis in his thigh. Alf therefore has two different concepts that he expresses by "arthritis" and "inflammation of the joints." But it would seem that the expert has only one concept expressed by the two expressions: for the expressions are synonymous, and the expert is expert. So it cannot be that Alf is able to deploy the same concept as the experts merely in virtue of his partial competence with the word and his deferential dispositions. If he did, then he would only have one concept, the unique concept expressed by both expressions in the experts' vernacular. So either Alf and the experts express different concepts by "arthritis" or by "inflammation of the joints" or both. Clearly the former is the most likely.

We have here the kernel of a good argument against consumerism. I will present the argument in two phases. Phase one continues to focus on synonymy, and merely generalizes the argument of the previous paragraph. Phase two reconstructs the argument in a way that dispenses with the appeal to synonymy.

Phase one

The basic problem generalizes to all twin cases based on incomplete information about a word's meaning. The generalization is not entirely straightforward, since some words lack precise synonyms. So, in order to generalize, we need to quantify over possible, non-actual synonyms. The idea is that in a case where an expression has no synonym, it still might have had one. For example, we could imagine that experts adopt a new word, say as an abbreviation, and use it explicitly as a synonym for the original. Or they might adopt a new word that derives from Greek or Latin roots, as part of a general revamping of their technical vocabulary. Again, this could be done with the explicit acknowledgement of synonymy with the original. If this possibility is accepted, then we can a generate the desired argument, as follows:

- Let w be the focal word.
- Let subj be the misinformed subject.
- Let c be the concept subj expresses by w.
- Let c' be the concept experts express by w.
- Let w' be a possible synonym for w.

The following counterfactual possibility exists:

- subj learns the correct, technical meaning of w', but he does it without learning its relation to w.
- c'' is the concept subj expresses by w'.
- c' is the concept experts express by w'.
- $c'' = c'$, since subj is expert in the use of w'.
- But c'' is not c, since subj has various beliefs involving c but not c'' and others involving c'' and not c.
- So, by the rules for identity, c is not c'.

In other words, c and c' differ in that the latter might come to be expressed by the new word, w', in a possible situation in which the former did not. But then different counterfactual possibilities are associated with c and c', and so they must be different concepts.[2]

Notice that the argument applies just as well to Putnam's "elm." To illustrate, let us take Putnam to be the subj, w to be "elm" and w' to be "ulmus" (the Latin name for elms). We could imagine that Putnam learns all about ulmi, but doesn't connect ulmi with "elm." Thus he doesn't believe that ulmi are elms. So he has two concepts expressed by the two words. But the experts express just one concept by both terms. Assuming that Putnam and the experts share a concept of ulmus, they must differ in the concept they associate with "elm."

A possible escape for the consumerist would be either to deny the possibility of true synonyms, or (what comes to much the same thing) to allow the existence of synonymy in some sense, but hold that synonymous expressions always express different concepts, even in the mouths of experts. Burge takes the latter option (Burge 1986).

Burge's claim that synonymous words are associated with different concepts is closely linked to his view that what a word means should be distinguished from the concept it expresses. Burge gives two arguments for this view. The first is that even experts may rationally doubt statements of synonymy. Burge illustrates with an example involving "sofa," but we can continue with "arthritis." An expert might come to doubt that "arthritis" just means *inflammation of the joints*. Anna, an expert, is contemplating the possibility of a strange new condition that causes intermittent ephemeral inflammations of the joints. These last for only a second or two. She considers whether this condition should be called "arthritis" and decides that it should not. We may suppose that Anna's view is not correct. "Arthritis" and "inflammation of the joints" are, in fact, synonyms. But in order to make sense of her thinking, we must attribute to her two different concepts associated with the two synonymous expressions. And so, Burge argues, we must distinguish linguistic meanings from associated concepts.

The second reason for making the distinction derives from the need to account for linguistic change. Suppose, for example, that Anna proves persuasive. She explains her reasoning to various doctors and lexicographers. They come round to her view. Received usage changes accordingly. Thus we want to allow for the following possibility: at one time, people thought that any inflammation of the joints was, by definition, arthritis, but at a later time, they came to think that only conditions involving non-ephemeral inflammations are arthritis. To make sense of this, we need to allow that the original concept of arthritis persists through the change in linguistic meaning. So, again, concept and meaning come apart. And, again, we must suppose that the concepts associated with the synonyms differ, since the experts doubt that arthritis is an inflammation of the joints only.

In order to block my argument, Burge would need to hold that synonyms can never express the same concept. For the argument hinged on the mere possibility of the target word's having an exact synonym that expresses the same concept as it does. So even if such cases are in reality extremely rare, the argument would still go through. And it does seem possible to construct plausible possibilities of the desired kind. Terms are sometimes introduced by explicit stipulation of synonymy. In such cases, it is hard to see how the synonymous expressions could express different concepts, since the introduced expression has no life of its own. It simply inherits all of its cognitive properties from its older sibling.

Something like this is actually taking place as I write. The Federative Committee on Anatomical Terminology (a sub-committee of the International Federation of Associations of Anatomists) is due to publish a new list of anatomical terms. The aims include standardizing vocabulary around the world, eliminating cumbersome expressions and replacing expressions that derive from the names of people with more informative terms. For example, "Fallopian tubes" will be replaced with "uterine tubes," "Eustacian tube" with (the much less cumbersome) "pharyngotympanic tube," and "Adam's apple" with "laryngeal prominence" (*The London and Manchester Guardian*, August 29, 1997, p. 4).

Suppose that the anatomists' idea of stipulating new terminology spreads to other branches of medicine. And suppose that there is a drive in particular for abbreviation and reduced expressions. A book listing the recommended usage is published and widely circulated. It contains the entry "Art $=_{df}$ arthritis." In this case, it seems that there is no possibility of the synonyms expressing different concepts. "Art" is just an alternative form of "arthritis." There is no way that the concept that experts associate with "art" could differ from the one they associate with "arthritis." The concept of art has no life of its own.

The original argument can now proceed as before. Alf comes to learn the word "art." He does not learn that it is an abbreviation for "arthritis." But he does learn that it applies, by definition, to all and only inflammations of the joints. By learning "art," he comes to have the concept *art*, the same concept that the experts associate with the term. The experts associate that same concept with "arthritis." But Alf associates a different concept with "arthritis," since he doubts the truth of "art is arthritis." So the concept Alf associates with "arthritis" is different from the concept the experts associate with the same word.

What goes for "arthritis" goes for other terms as well. The same or similar sorts of possibility arise: experts in the relevant domain adopt an explicit policy of adopting a new word as a precise synonym of an existing one. In such cases, no concept is available to be associated with the new term except the one associated with the old.

Notice also that Burge's attitude to synonymy introduces a slight structural weakness into his overall position. As we saw, Burge's desire to distinguish the concepts associated with synonymous expressions led him to distinguish also between the concept associated with an expression and the expression's meaning. But this latter distinction undermines part of motivation for the first step of the original thought experiment.

Let us return to Alf, at the stage prior to his linguistic enlightenment. Suppose he says, "I have arthritis in my wrists." One of the thoughts driving the original externalist intuition was that we should credit Alf with the concept of arthritis because of his minimal competence with the word and his deferential dispositions. But once word meaning and associated concept are distinguished, this thought loses force. For we may interpret his words by assigning them their conventional meanings, as the externalist would recommend. But the meaning his words express is distinct from the content of his belief. Thus we need not suppose that his belief has the same content as the belief that an expert would express using the same words.

To put the point slightly differently: the externalist can no longer bolster the first step of the thought experiment by claiming that what Alf says is what he believes. For the externalist should take 'what he says' to be the conventional meaning of the words he utters. (Otherwise the connection between Alf and conventional meaning is broken.) But then what he says is not what he believes, for concepts and word meanings are to be distinguished. And, again, if what he says is not what he believes, then why should we suppose that what he believes is the same as what the expert believes? Why should we think that the concept he associates with "arthritis" is the same as the experts'? It seems to make more sense to deny this, and hold that his concept is not the same as the expert's. After all, Alf's concept differs from the expert's in a fundamental aspect of its psychological role. For Alf is willing to believe that he has arthritis in the thigh. But the expert is unwilling to believe this, on conceptual grounds alone, without resort to any empirical investigation.

There is a further reason to suppose that it is at least possible for synonyms to express the same concept. I take it that most externalists would concede that facts of the form *word w expresses concept c* have supervenience bases of some kind or other. But then, exactly similar supervenience bases could underlie the relation of a single concept to two words: whatever base underlies the fact that one word, *w*, expresses concept *c* could be duplicated with respect to another word *w'*. Suppose, for example, that there are two almost identical twin rheumatologists called Mike and Dan. Both, of course, are experts about arthritis. Mike uses the abbreviation "art" to express this concept, while Dan uses the term "arthritis." But that is the only difference. The beliefs, dispositions, etc., that Mike associates with "art" are identical to those Dan associates with "arthritis." Here, it would seem that whatever it is that makes it the case that Mike expresses his concept by "arthritis" is duplicated with respect to Dan and the concept he expresses by "art."

The externalist might be tempted to bite the bullet for Burgean reasons. He might hold that Mike and Dan could, upon reflection, come to doubt that art is arthritis. This possibility alone shows that "art" and "arthritis" express different concepts after all. I don't think this is very plausible in the case described. As matters stand, they have no reason to formulate this doubt since they know full well that "art" merely abbreviates "arthritis." It is true, however, that Mike and Dan might come to conceive of a lay subject, like Alf, who doesn't know that art is arthritis. So it seems that they can reasonably formulate the second order belief that someone else might doubt that art is arthritis.[3] And this could motivate the idea that "art" and "arthritis" express different concepts for the experts. I don't think that is quite right, for reasons I will come to later. But however that strand of the dialectic goes, it is possible to reformulate the original argument in a way that avoids the issue. I proceed to do this in phase two.

Phase two

The core of the argument is the claim that there could be two words that the misinformed subject uses to express different concepts, but that express just one concept of the expert's. Cases like this can be constructed involving a unique expert who does not even have the relevant words in her repertoire. There are two words in the misinformed subject's vocabulary, but only one in the expert's. Let us look at some examples.

Mary is a consultant rheumatologist. In 1999 she writes a definitive article about arthritis. In 2005 she unfortunately dies. In 2010, two separate publishers publish the article in two separate volumes. One uses the newly introduced abbreviation "art," the other uses the venerable "arthritis." Bert happens rather casually to browse the two volumes at different times. He comes to believe that "art" applies to all and only inflammations of the joints and "arthritis" applies to any disorder in which aches and pains affect muscles and joints. He believes, as he would say, that he has arthritis in his thigh. But he doesn't believe that he has art in his thigh. So he expresses two concepts by the two terms. But surely Mary has just the one concept that comes to be expressed by the two terms.

Perhaps the externalist would reply that, in this case, other experts, the ones to whom Bert would defer in 2010, might come to doubt that art is arthritis. So the fact that Mary had only of the two concepts is not relevant. This reply does not make much headway, since one could equally suppose that there are no experts left in 2010, so all deferential relations lead back to Mary. If this is not accepted, the next examples avoid the problem anyway.

Mary is a consultant rheumatologist. In 1999 she writes a definitive article about arthritis. In 2005 she unfortunately dies. In 2010 a publisher publishes it in a volume. In 2010 Bert happens to browse the volume and as a result forms the belief that "arthritis" applies to any disorder in which aches and pains affect muscles and joints. In 2015 he happens to come across a rather inferior photocopy of a section of Mary's article and begins to browse. His eye-sight is not very good, so the *r*'s and *n*'s in this inferior photocopy look rather similar to him. He thinks he is reading about a condition called "anthnitis." He reads a crucial paragraph in the article that defines "arthritis" and comes to believe that "anthnitis" just means$_{df}$ *inflamma-tion of the joints*. If asked, he would say, "Anthnitis is not arthritis."

In this example, it is surely correct to say that the experts have just one concept where the subject has two. For no expert has the word "anthnitis" in her repertoire. There are two words in Bert's repertoire, but the deferen-tial paths associated with both of them lead back to the experts' unique term "arthritis" and the unique concept it expresses.

Any attempt to pursue the Mates considerations at this point leads to absurdity. The idea would have to be that an expert who has not actually come across the term "anthnitis," who has never heard or thought of it, is nev-ertheless such that he might, counterfactually, come across the term "anthnitis" and might then come to think that someone doubts that arthritis is anthnitis. And merely be-cause of this, we have to say that the expert already has two concepts expressed by the two words. But that is absurd. It might be reasonable to hold that after an expert acquires the term "anthnitis," she might then form a corre-sponding concept differing from her concept of arthritis. But it is not reasonable to suppose that she has the two con-cepts before she has heard or thought of the word "anth-nitis"! A few more examples should make the point vivid.

Kripke's Pierre believed that Paderewski (the musician) had musical talent and that Paderewski (the statesman) lacked musical talent (Kripke 1979). I once knew someone who believed that Robert Nozick (the political philosopher) was not a philosopher of science but that Robert Nozick (the philosopher of science) was a philosopher of science. In fact, Paderewski is Paderewski and Robert Nozick is Robert Nozick. What goes for proper names goes for general terms as well.

Many general terms in fact have two meanings one of which is a generalization of the other. For example "psychopathy" can either apply to mental illness in general or it can apply to a specific form of social dysfunction involving lack of conscience. "Tea" as applied to drinks can either denote infusions of tea leaves or, more generally, infusions of various fragrant leaves. "Grog" in Australian English can either denote any alcoholic drink, or a specific kind of drink made from spirits. Given this, it easy to imagine someone who falsely believes that a certain word form has two related but different meanings. Generalizing from "tea," for example, Cath might believe that "coffee" can either mean a drink made from ground coffee beans, or any dark, bitter tasting drink. She might then believe that not all coffee (general sense) is coffee (specific sense). Don might believe (incorrectly) that "optician" can either denote any specialist in optics, or, more specifically, a specialist in optics who is commercially involved in the correction of people's sight. Thus Don believes that not all opticians (general sense) are opticians (specific sense). And Bert might believe that "arthritis" can denote either a family of diseases causing inflammations of the joints, or, more generally, any rheumatoid condition involving aches and pains of the joints, muscles, tendons or ligaments. Thus Bert believes that not all arthritis is arthritis.

The same result can arise when one kind manifests itself in different ways (like the individuals Paderewski and Nozick). Observing some tadpoles, Frank wonders what those little fish are. He is told that they are young frogs. He comes to believe that "frog" is ambiguous between a term for a kind of fish and a term for a kind of amphibian. He thus has two concepts where the experts have one. Ginny is a first year undergraduate in philosophy. She believes that "internalism" is ambiguous. In one sense it denotes a plausible thesis about the relation of psychological states to the brain. In the other sense it denotes a radical form of idealism, holding that the world itself is internal to the subject's mind.[4] Harry, having come across tuna in both live and culinary forms, comes to believe that "tuna" is ambiguous between a kind of fish and a kind of aquatic mammal (as "dolphin" is in fact).

All of these examples could be spelled out so that the subjects involved had the kind of partial understanding and deferential dispositions that would, according to the consumerist, allow them to express the experts' concept by their use of words. In each case the subjects have two words that express different concepts of theirs. These two words are at the ends of deferential paths that lead back to just one concept of the experts. If one were to deny this, then one would be committed to the claim that any time any subject suffered a misunderstanding of the relevant kind, all the experts (hence all the rest of us) would immediately acquire a new concept. This is not an acceptable result.

3.3   DIAGNOSES

I have argued that the social externalist does not draw the right conclusions from Putnam's and Burge's examples.

Nevertheless, the intuitions underlying the view can seem compelling. I think that there are two different kinds of intuition at work here, a first personal one and a third personal one. I think that both can be diagnosed. We begin with the former.

The first person intuition centers on the manner in which the ill-informed protagonists accept correction. After Alf has conversed with the doctor, he regards himself as having previously believed that he had arthritis in the thigh. This goes against my internalist claim that Alf did not, prior to the conversation, have the concept of arthritis. I suggest that to understand Alf's view of the matter, we need to think of concepts as organic entities that can persist through changes of extension. Alf takes it that after correction he still deploys the same concept he had earlier. In a sense he is perfectly correct. It is the same concept in the sense that it is the same organic unity that has survived the conversation with the doctor. However, it has undergone a change of cognitive content and even of extension-conditions.

Post Fregean philosophers might find this use of "concept" strange or unacceptable. But really the matter is just terminological. Developmental psychologists frequently discuss the natural evolution of a concept over time. For example, a number of psychologists studying the development of mentalistic concepts (concepts of belief, desire, perception, etc.) hold that typical three-year-olds do not have the adult concept of belief. Rather, they possess a simplified concept that represents states that are something like copies of real situations in people's minds, or perhaps nonrepresentational states that relate people directly to situations in the world. (See, e.g., Gopnik and Wellman 1992, Perner 1991, Wellman 1990.) Call the original concept *s-belief* ("s" for simple) and the mature concept *belief*. The s-belief concept evolves into the belief concept. It doesn't matter whether we describe this as one concept changing its semantics over time or as an ancestor concept ceasing to be and being replaced by a descendent concept. The point is that there is a natural process that involves some kind of mental entity persisting through a semantic change. We could choose to call this evolving mental entity a concept.

The third person intuition is simply that we find it natural to say of Alf that he believes he has arthritis in his thigh. To understand this, it will help to consider some more (brief) examples. Alf says "I have a hippopotamus in my fridge." We incline to the view that Alf believes he has a hippopotamus in his fridge. But he goes on: "It is approximately the size and shape of a tennis ball, has an orange, wrinkly skin, and moist flesh with a pleasant, tart flavor." We revise our view, and decide that when he says "hippopotamus" he means *orange*. (The example is from Davidson 1984, 100–101.)

Fred says, "I have rheumatism in my wrists." We incline to the view that he believes he has rheumatism in his wrists. But he goes on: "I used to work in a factory, and was constantly having to rotate my hands. This caused an inflammation of the joints. 'Rheumatism' just means *inflammation of the joints*, you know." We revise our view, and conclude that by "rheumatism" he means *arthritis*.

The first case illustrates the uncontroversial point that we are not always averse to reinterpreting a subject's words. Alf says "hippopotamus." Once we learn that the concept he thereby expresses is expressed by our word "orange," we use the latter in our ascriptions of contents to him. The second case illustrates that we are not averse to doing this in cases where the word the subject has chosen does not mean exactly what he thinks it does, but still has a closely related meaning. In both cases, the subject misunderstands the word he uses. But the concept he associates with it is one that we happen to know and that we can express by a single word of our language.

Burge's original subject says, "My grandfather had arthritis. Now I have it in my hands and wrists. My joints get inflamed. It is painful." We incline to the view that Alf thinks his grandfather had arthritis, etc. Then he says "My arthritis may have spread to my thigh." Again, the subject misunderstands the word he is using. But now we do not find it so natural to reinterpret his words. Rather, we incline to say that he believes he has arthritis in his thigh. Why is this?

Two features appear to distinguish this case from the previous one. First, we don't know exactly what Alf takes "arthritis" to mean. Since we don't know that, we are not well situated to find an expression that means in our mouths what Alf takes "arthritis" to mean in his. Second, given what we do know, no relatively short expression, no word or simple phrase, springs to mind as a good candidate for the reinterpretation. By contrast, in the "rheumatism" example, Alf conveniently provided us with an account of what he takes the word to mean. And, it so happened, that very account coincided with the definition of a single word in our vocabulary: "arthritis." Thus we knew what content we needed to capture and we had a convenient word for capturing it.

Another example should help us tease apart these two factors. This time Alf says, "I have arthritis in my thigh. Arthritis is a specific kind of rheumatism. It has something to do with the auto-immune system. White blood cells attack the bones and muscles. It is a hereditary condition. There is no known cure." Now we know a fair amount about what Alf takes "arthritis" to mean. How do we report what he thinks he has in his thigh? My feeling is that it is no longer natural to say "He thinks he has arthritis in his thigh." In this case our minds are focused on the nature of Alf's idiosyncratic conception and on the difference between that and our concept of arthritis. On the other hand, we also find ourselves unwilling or unable to reinterpret. We have no word or short phrase that captures Alf's idiolectic meaning. There is the possibility of something like "He thinks he has in his thigh a certain kind of hereditary auto-immune disease involving white blood-cells attacking bones and muscles." But only a pedant would say that.

Thus, although we find it acceptable to say "Alf believes he has arthritis in his thigh" it does not follow that Alf has the concept of arthritis. He doesn't have the concept—our concept, the expert concept—of arthritis. Rather, our acceptance of that report is due to two factors. First, we do not know in any detail what his concept is like. Hence we are not forced to take notice of the precise nature of his concept and how it differs from our concept of arthritis. So we are not averse to accepting "arthritis" in our reports of his beliefs. Second, we don't have a ready alternative word or short phrase by which to reinterpret. So even if we were not happy to stick with "arthritis," we would not have any better option.

The diagnosis is thus that consumerism misconstructs an artefact of our practice of attitude reporting. Since we are not conscious of all of the principles underlying that practice, this misconstruction is understandable. However, the first two examples of this section indicate that it really is a misconstruction. We do reinterpret the subject's words in order to find the best match between his concepts and our vocabulary, when we know how to do so. Given this, our failure to reinterpret and our willingness to use the subject's own words in our reports of his attitudes in Burgean examples is predictable from the factors discussed. The alternative, after all, would be not to attempt to report on the subject's attitudes at all. And it is no surprise that we are not drawn to that option.

On a plausible account of belief reporting, "Alf believes he has arthritis in his thigh" could actually be true, even if Alf lacked the concept of arthritis. The idea is this. When someone reports on a belief, he uses a sentence of his own language as a sample representation. It is very much as if one holds up a sentence of one's own, S, and says, "S: that's what he believes." Very roughly, a report of the form "a believes that S" as uttered by b, in conversational context C, is true iff the content of S in b's mouth is similar enough, by the standards of C, to some belief of a's. "Similar enough" does not mean *identical*. And standards of similarity shift considerably from context to context. So, in ordinary speech it may be perfectly correct to say of the misinformed Alf that he believes that he has arthritis in his thigh, even if he has no concept that is exactly that of arthritis. This is particularly likely if we have no exact idea of Alf's concept and we lack a convenient short phrase of our language to express it. However, in a stricter context, one where we are interested in the precise details of Alf's psychology, such a report would be incorrect.[5]

The metalinguistic account of belief reporting also helps explain some tricky situations that have arisen. Burge himself presumably believes that Alf believes that arthritis is not an inflammation of the joints. On the other hand, it is likely that Burge doubts that Alf believes that an inflammation of the joints is not an inflammation of the joints. But if I am right, then "arthritis" and "inflammation of the joints" express the same concept of Burge's. And this makes Burge seem incoherent, which of course he is not. The meta-linguistic account explains this. Since "arthritis" and "inflammation of the joints" are different linguistic items and belief reports relate subjects to linguistic items, Burge's second order belief is perfectly coherent.[6]

Now, recall Pierre, who believes that Paderewski isn't Paderewski. It can appear that what I just wrote makes no sense. "Paderewski" is a single name with a single meaning. So when I use it to talk about Pierre's belief, it seems that I commit Pierre to a logical incoherence of the most obvious sort. But that is not the desired effect. Pierre's condition is clear enough. He has two names in his idiolect that happen to be spelled and pronounced the same. These are "Paderewski$_1$" and "Paderewski$_2$"—just as we have "bank$_1$" and "bank$_2$." "Paderewski$_1$" and "Paderewski$_2$" have different senses in Pierre's mouth, he associates different concepts with them. "Paderewski$_1$" names a musician and "Paderewski$_2$" names a statesman. In order for me to report on Pierre, I need temporarily to adopt his language. I too must adopt the two names for the purpose of talking about Pierre. Having done so, I can write, coherently enough: "Pierre believes that Paderewski$_1$ isn't Paderewski$_2$." In ordinary expression, however, we don't pronounce or write subscripts. That is why it is hard for us to express what we want to. We have to settle for something like "Pierre believes that Paderewski (the musician) isn't Paderewski (the statesman)."

Now we can understand how "Alf believes that arthritis is not an inflammation of the joints" can be true, while "Alf believes that an inflammation of the joints is not an inflammation of the joints" is false. Since we can make sense of these reports, it seems that we must associate different concepts with "arthritis" and "inflammation of the joints." That is essentially the Mates/Burge inspired reason for saying that synonyms never express the same concepts, even in the mouths of experts. But I suggest that what is really going is this. "Arthritis is not an inflammation of the joints" and "An inflammation of the joints is not an inflammation of the joints" normally express the same thought for us. But they can cease to so in the specific context of a belief report. In that context, the words are used as sample representations for the purpose of representing what someone else believes. When used as samples, the words need not retain their normal senses. As with Pierre, so with Alf. When we use the terms in the belief report, we ask our audience to interpret the words not as usual, but in a way that makes sense of the reportee. At that point, in the context of talking about Alf, our synonymous expressions cease to be synonymous and are used to express different concepts on our part. But these are really just Alf's concepts that we have adopted for the purposes of understanding him.

So when I say (as I indeed believe), "Nobody could believe that arthritis is not inflammation of the joints," I say something true. For I am using the words as I normally use them. Nothing has arranged the context so that I can use them with someone else's senses. But were Burge to say "Alf believes that arthritis is not an inflammation of the joints" he might speak truly, using "arthritis" in Alf's way.[7]

3.4   A WEAKER CONSUMERISM

The arguments are supposed to have shown the cognitive content of a deferential concept does not derive from the views of the experts to which the subject defers. That is to say, mere partial competence with a term together with appropriate deferential dispositions do not allow subjects to deploy the same concepts as the experts.

The consumerist might adopt a weaker thesis (see Burge 1989). He might concede that the cognitive content

DEFERENTIAL DISPOSITIONS AND COGNITIVE CONTENT

of a deferential concept is not completely fixed by the experts. Thus Alf does not have the same "arthritis" concept as the experts. But still, the extension of Alf's concept is fixed by the experts. And, given that a difference of extension yields a difference of concept, Alf and Twin Alf have different "arthritis" concepts.

Even if the extensions of Alf's and Twin Alf's concepts are different, it is natural to think that they have the same cognitive content. For they have exactly the same cognitive roles: they feature in otherwise identical desires, beliefs and so on. And these desires, beliefs and so on themselves play exactly similar roles in the twins' psychologies, in the way they think and reason and so on. What this shared cognitive content might consist in and how we might talk about it are topics for the next two chapters.

However, I think there is some reason to doubt that Alf's concept has even the same extension as ours. Alf does not express our concept of arthritis when he says "arthritis." So, if we judge it right to say that Alf believes he has arthritis in his thigh, we are not using the term "arthritis" in the standard way. Rather, we use it to express Alf's concept. Our willingness to reinterpret when we have the knowledge and means to do so suggests that in these special cases, when we use terms to capture the concepts of others in attitude reporting, we may not be using them with their normal extensions. For example, in the case where Fred said "I have rheumatism in my wrists. 'Rheumatism' just means *inflammation of the joints*," we are happy to say that he believes he has arthritis in his wrists. In so doing, we use "arthritis" rather than "rheumatism" to report on his belief, even 'though these terms have different extensions. It may well be, then, that when we use the Alf's term, "arthritis," to express his concept rather then ours, we are using it with a non-standard extension.

Segue

I will assume from now on that cognitive content is narrow. I have not explicitly argued against every possible version of externalism, claiming that cognitive content depends in this, that or the other way upon relations to this, that or the other social or nonsocial external matter. But I have argued against the most popular and influential ones. Rather than try to refute every externalist position, I turn now to other aspects of the debate.

The notion of cognitive content I have deployed is as yet unexplicated. All I have assumed in this chapter is that if the cognitive role of concepts differs, then so does the cognitive content. If a subject has a belief involving a concept c, but not involving a concept c', then c differs in content from c'. This use fits with the way "cognitive content" featured in chapter one. There, the cognitive content of a concept was held to be what accounts for the concept's role in psychological explanation. So, for example, the "water" concepts of twinned Bedouins on Dry Earth, XYZ Twin Earth and Earth share a cognitive content because a good psychology would subsume all of them under the same generalizations.

The remaining positions are two factor theories and radical internalism. Two factor theories concede externalist claims about extension conditions, but propose an additional kind or level of narrow content. Radical internalism denies the externalist claims about extension conditions. I have not yet said enough to select among these alternatives. That is the business of the next two chapters. Chapter four examines and rejects the chief two-factor accounts. Chapter five motivates and defends radical internalism.

4 *Cognitive Content and Extension*

The preceding chapters have promoted the idea that concepts have narrow cognitive contents. If the idea is right, then twins' counterpart concepts, such as Oscar's and Twin Oscar's "water" concepts and Alf and Twin Alf's "arthritis" concepts, share fundamental psychological properties. They should be classified together by a good psychological theory. But, beyond that, not much has been said about the nature of cognitive contents or the way in which a psychological theory could describe them. The internalist arguments of the preceding chapters need to be accompanied by some positive account of narrow cognitive content and the way it is drawn upon in psychological explanation.

Most internalists have tended to favor two factor accounts of content. Two factor accounts accept an externalist view of extension conditions. Either they accept the Putnam-inspired account of the extension of kind concepts, holding, for example, that Oscar's and Twin Oscar's "water" concepts have different extension conditions. Or they accept the Putnam- and Burge-inspired account of the extension conditions of deferential concepts, agreeing, for example, that Alf and Twin Alf's "arthritis" concepts differ in extension conditions. Or they may accept both. But they also accept the internalist thought that twins' concepts have some kind of shared, extension-independent cognitive content. They thus recognize both wide and narrow content.

This chapter assesses and rejects what I take to be the leading two factor accounts. These are functionalist accounts, character accounts and descriptive accounts. In the next chapter, I will offer my own account of narrow content.

4.1   NARROW FUNCTIONALISM

Functionalism is the thesis that psychological properties can be identified with properties specified in terms of actual or potential causal relations. Suppose, for example, that Fred believes that there is an alligator in front of him. This belief has a characteristic set of causal properties. It is the kind of belief that would typically be caused by the presence of an alligator appropriately situated relative to Fred. If Fred fears alligators, then the belief will cause him to be afraid. And the belief and the fear combined will cause him to take evasive action (supposing that he doesn't get rooted to the spot). On the other hand, if he believes he is hunting alligators and is armed with an alligator knife, he might be motivated to approach the beast with a view to a kill. And so on. The idea behind functionalism is that such causal relations exhaust the nature of the belief that there's alligator in front of one. All and only states with that characteristic causal role are beliefs that there's an alligator in front of one.

A standard way to develop the idea, deriving from the work of F. P. Ramsey (1931) and David Lewis (1970, 1972), goes as follows. Take your favorite content-using psychological theory, e.g. common-sense or cognitive psychology or some combination of the two. The theory specifies causal relations among (i) inputs (ii) psychological states (iii) outputs. Leave your input and output terms in place, but remove all the psychological terms and replace them systematically with variables. Each variable now marks a state with a specific type of causal role, the role being specified purely in terms of causal relations to inputs, outputs and other states. Each variable thus marks the functional role associated with the psychological state picked out by the psychological term it replaces. Here is a more technical version.

Let T be a psychological theory specifying causal relations among psychological states, inputs and outputs, respectively:

$$T(s_1, \ldots, s_n, i_1, \ldots, i_k, o_1, \ldots, o_m).$$

The "Ramsey sentence" of T redefines $s_1, \ldots, s_n$ in terms of causal relations among the states, inputs and outputs, and existentially quantifies over these states:

$$\exists x_1, \ldots, x_n [T(x_1, \ldots, x_n, i_1, \ldots, i_k, o_1, \ldots, o_m)]$$

We now define the "Ramsey functional correlate" of s_i:

$$\lambda y \exists x_1, \ldots, x_n [T(x_1, \ldots, x_n, i_1, \ldots, i_k, o_1, \ldots, o_m) \ \& \ y \text{ is in } x_i]$$

"λ" is called "the property abstractor." The expression "$\lambda y F y$" is read as "the property of being F" or "the property y has iff y is F." So the Ramsey functional correlate of s_i is the (second order) property of being in some state x_i, as defined by the Ramsey sentence of T.[1]

In its most straightforward version, functionalism holds that psychological states are their Ramsey functional correlates. On this version, if one thinks that the chosen psychological theory, T, ascribes wide contents, then the functional states must be wide as well. The definition of the functional states must then include some wide factors. This could be effected by including external matters among the inputs and outputs and adding a requirement that the subject actually causally interact with these, or by relativizing the roles to particular environments (as in: to believe that water is good for plants is to be in a functional state of the appropriate type in an H_2O environment). This is the line taken by externalist functionalists (e.g., Harman 1987).

There are two ways one could use Ramsey's technique to get an internalist version of functionalism. One way would be to select an internalist psychological theory, one that already treats twins identically, as T. This of course raises the question of what an internalist psychology would be like. I think this is exactly the right question for the internalist to address, and I will return to it in the next chapter. But this version of functionalism then merely assumes that some kind of psychology will give us narrow content, and then adds the further claim that narrow content reduces to functional role. This further claim is of no direct concern to the dialectic between internalism and externalism.

The other way is to reject the identification of functional states with psychological states attributed by T and replace it with an identification of functional states with narrow psychological states. The idea is to accept that psychology is, on the face of it, externalist and would ascribe different contents to twins. It would say, for example, that Oscar believes that water is good for plants, while Twin Oscar believes that twater is good for plants. However, the idea is, a psychological theory would assign these two states the same functional role, the same pattern of potential causal interactions. Oscar's water beliefs would be caused by, and would in turn also cause, exactly the same sorts of things as Twin Oscar's twater beliefs. Thus we would take T to be a wide psychological theory, and use Ramsey's technique to abstract a narrow one from it. The Ramsey sentence of T, in effect, becomes narrow psychology, classifying psychological states purely in terms of their causal potentialities. Narrow contents derive from these causal potentialities.[2]

The original wide contents of T are reduced to some kind of combination of the narrow functional roles together with extension conditions, the latter being accounted for by a separate reductive theory, probably a causal one.

I will call this version of functionalism "DRNF$_1$" for "directly reductive narrow functionalism, version one." It is narrow functionalism because it is functionalist about narrow content. It is directly reductive in that it gives us an account of what narrow content reduces to before giving us an account narrow content. And it is version one because I will discuss a second version later. Let me elaborate.

Normally, a reductive theory works by relating the vocabularies of two theories or two sets of generalizations: it says that phenomena picked out by one of these vocabularies are the very same as those picked out by the other. For example, the quantities picked out by thermodynamics can be identified with quantities picked out by statistical mechanics. DRNF$_1$ offers a reductive account of narrow content in the sense that the vocabulary in which the functional roles are specified is free of intentional terms like "reference," "extension," "content" and so on.

The account is directly reductive in the sense that this is the only theoretical vocabulary it gives us with which to frame generalizations in terms of narrow content. It doesn't provide a theory at the unreduced psychological level in which explanatory and predictive generalizations about narrow content can be framed. It moves directly from a theory that talks about wide contents to the functional roles to which narrow contents are reduced. The best we can do by way of an unreduced vocabulary, I suppose, is this sort of thing: "the narrow content of the belief that fish swim," or "the narrow content of Oscar's belief that fish swim" or something along those lines. Internalist philosophers will claim to understand such terms. But these are not the terms of an existing explanatory and predictive theory.[3]

I have two qualms about DRNF$_1$. One is that it might be false. The other is that, because of its directly reductive nature, it leaves important questions unanswered. I begin with the former.

System relativity and infinitary theories

The causal role of a psychological state transcends its tendencies to interact with inputs, outputs and other psychological states. The nature of its interactions depends also on various features of the cognitive system to which it belongs. Here is a simple example. Alf and Bert are pretty much identical in respect of their outlook on the world, their beliefs, desires, hopes, fears and so on. Alf and Bert are severally confronted by an enraged gorilla. This input causes both of them to think *There's an enraged gorilla right in front of me*. Both of them get very frightened. Alf backs away slowly, saying "easy boy, easy." But Bert remains rooted to the spot, frozen in fear.

This difference between the effects of the belief across the two subjects does not stem from any difference in their psychological states per se. They both fear enraged gorillas. And we may suppose that, to the extent that degrees of fear are quantifiable, they fear enraged gorillas to the same degree. The difference is just that Bert's freezing up mechanism is more sensitive than Alf's.

There are many kinds of systemic features that have a similar sort of impact. Freud's Censor mechanism operates on, roughly speaking, the distress-generating potential of a psychological state. The more unbearable you find it that you want to kill your father, the harder the Censor will work to repress your murderous Oedipal desire. Again, we can imagine two subjects who share all their desires, beliefs and so on, but whose Censors differ in their sensitivity. The shared psychological states will have different functional roles in these subjects.

Memory buffer space is another obvious example. Alf and Bert may both believe p, q, and r and may both be considering whether s follows. Alf figures out that it does. But Bert cannot perform the inference. He loses track in the middle because it is too complicated for him.

Another, more global property of cognitive structure that has the same effect is a consequence of modularity. There is good evidence that minds are modular, at least to some extent. Even the weakest notion of modularity has it that there are information filters within the mind.[4] This idea goes back to Freud: what is in the unconscious is not, or not easily, accessible by the conscious mind. So we could imagine two people who have beliefs and desires with the same cognitive contents, but which differ in their distribution across the conscious/unconscious divide. Given the same inputs, these people will produce different outputs, because of the different patterns of intra-mental interaction they are subject to.

Current linguistics and cognitive psychology describe many more such divisions: information in the language faculty is not available to the conscious mind, nor is it available to the visual system. Information in the visual system is not available to the conscious mind, nor, vice versa, is information in the conscious mind available to the visual system. So, again, it is possible to envisage creatures of different kinds who are matched in respect of the contents of their psychological states, but differ in their cognitive dynamics because the information filters are differently organized: in one, certain items of information can combine, where in the other they cannot.

All of this poses a serious problem for DRNF$_1$. Consider Alf and Bert. They both believe that there is enraged gorilla in front of them. This belief has the same cognitive content in both subjects. But, due to the different sensitivities of their freezing up mechanisms, the belief has different effects in the two subjects. So, to find a common functional role for Alf's and Bert's belief, we need a theory that can tell us not only how the belief interacts with other psychological states, inputs and outputs, but also how these interactions are themselves affected by features of the psychological mechanisms in which they are embedded. In the particular case of Alf and Bert, this might be possible. But it is far from clear that any psychological theory would be sufficiently encompassing to capture all pertinent systemic features.

In order to specify all of the relevant functional properties of a given psychological state, we would have to specify its role in every possible cognitive system. Given all the variables involved, it is likely that we would need to specify its role in infinitely many different kinds of system. And it is far from evident that any finite theory would be able to do this. Current psychological theories specify the behavior of types of representation within the specific cognitive systems they study. But they do not appear to provide a way of generalizing to other systems. Theories of human vision, for example, tell us about the functional roles of representations of edges, surfaces and discontinuities in human visual systems. But they do not describe these in a way that would generalize to other, differently organized visual systems. It is not clear that the final, complete, true psychology in Platonic heaven is any different. There may simply be no finitely specifiable functional role associated with each psychological state.[5]

Perhaps the final psychology contains infinitary specifications of the role of each psychological state in every possible type of cognitive system. Perhaps we could identify contents with these infinitary roles. I have no proper objection to this proposal. But unless a good argument is provided in its favor, it seems inelegant enough to warrant scepticism. Moreover, there are related but more general grounds for discontent.

An alternative metaphysics

There is another objection to functionalism that would, if correct, help to explain some of the awkward consequences discussed in the previous subsection. Token objects, events, processes or instantiations of states are the items that play the causal roles associated with psychological properties. When Alf sees the gorilla, certain events or processes ensue in his brain that provide the neurological realization of his belief that there's an enraged gorilla in front of him. And this token belief, in its particular context, causes him to back away. It is natural to think that the reason the belief has this effect is because it has the content it has. The psychological properties endow it with its causal powers. But if that is right, then the psychological properties cannot simply be those causal powers.

If we identify the psychological property with the causal role then we lose the obvious explanation of why the event has the causal role it has. But then either something else explains the causal role, or nothing does. And neither of these options seems attractive. If there is some other explanation of why, say, the belief causes the backing off, then it would be neurological or syntactic-computational or otherwise lower-level. The problem with this is that most considerations that support a functional role account of psychological properties work just as well for neurological and other lower-level properties, as many functionalists would agree (see, e.g., Lycan 1987). Then we get into a vicious regress: for we need to explain why synapse firings have their causal roles, and we drop down to yet lower-level electro-chemical properties. We go functionalist about them. And we are off on a path to a functional role account of the properties of sub-atomic particles. We end up on the other horn of the dilemma, holding that token objects and events have causal powers, but these are never ultimately explicable.[6]

If we go functionalist about all properties, or all natural properties, then we can explain why some event has the causal role it has by pointing to the causal roles of some of its constituents. But we must accept that explanations eventually run out. Either there is a bottom level at which we accept unexplained causal roles (it is just a brute fact that quarks do this, that and the other: end of story) or there is an infinite regress. Again, I have no proper objection to this view. But it seems unnecessarily mysterious. Why not take the world at face value and accept that objects and events do what they do because they have the properties they have? Accepting this allows for an attractive metaphysical position.

Some properties endow their possessors with causal powers. We can think of causal powers in terms of a schema: in circumstances C, given input I, object O will have effect E, ceteris paribus. If I drop my key, it will fall to the ground. The circumstances include the presence of the Earth nearby and the absence of interfering factors. It falls because it has mass and the Earth has mass and massive bodies attract one another. Had there been a powerful magnet above, the key would have risen. Then its mass would not have caused it to fall, but would have slowed its ascent. The mass of the key explains one causal disposition in one set of circumstances and another in another.

The problems with identifying the mass of the key with some specific set of causal powers mirror the problems with identifying psychological properties with functional ones. Massive objects do different things in different physical systems. As far as we know, any statement of what the mass of an object would cause it to do in certain conditions would fall short of the whole story. There might be further conditions to consider. And the ceteris paribus clauses always have to remain in place, which should tell us that there is more to having a given mass than is specified by any theory of its causal role.[7]

I am not offering the above picture as any sort of panacea. The metaphysics of properties and causation remains deeply perplexing. Even if the picture is right, a descendent of the old problem remains.[8] The mass of an object causes it to fall to Earth or impedes its progress towards a magnet. But why? Here we are back with the problem of entering a regress of explanations of the same type or just accepting some brute facts. In the first case we move to explanations of the form "Having mass M is associated with having properties p, q, and r, and these cause an object to do x, y, and z in circumstances C." In the second case, we say, "Having mass M just does endow an object with the disposition to x, y, and z in C full stop."

But still it seems more plausible that, in the final analysis, an object's or event's causal powers are owed to some of its properties rather than to nothing. It is not just a brute fact that an object or event tends to cause what it does. Or so it seems. Further, as I said, the picture underwrites the initial objection to functionalism. For if the properties in question endow their possessors with causal powers rather than merely being causal powers, then it comes as no surprise that it seems impossible to give a specification of exactly those causal powers that are associated with a target property. Properties interact with each other. What an object with a given property does, depends upon what it interacts with: the embedding conditions and the other denizens of those conditions. Since there are indefinitely many types of embedding conditions and fellow denizens, there does not seem to be any principled way of exhaustively stating the causal powers associated with a given property.

Be all that as it may. The real problem with $DRNF_1$ as an account of narrow content in the present context is that it fails to address pressing questions that need to be addressed.

$DRNF_1$ and narrow content

As I mentioned above, any good arguments that psychological properties are functional should extend to all or at least many other properties; computational-syntactic ones, neurological ones, chemical ones and so on. A unified treatment of natural properties seems far more attractive than one that treats the psychological properties, but not, say, neurological and chemical properties, as functional. Functionalism then counts as a variety of naturalism about the mind. The point of the theory would be to head off Cartesian and certain other versions of dualism by saying: psychological properties are like other natural properties; all such properties are functional.

But the question that the internalist has to face is: how are we to frame generalizations in terms of narrow properties? The picture offered by $DRNF_1$ is something like this. First, take your best psychological theory. Then construct its Ramsey sentence. We can now introduce names for the functional states it defines by replacing the bound variables with constants, say $F_1, \ldots, F_n$. We then frame our narrow generalizations in terms of these: anyone in states F_i, F_j, F_k, given input I_k, will produce output O_n.

This proposal might work for Oscar and his twin. T could be something like common-sense psychology with externalist content ascriptions. T would then ascribe water beliefs to Oscar and twater beliefs to Twin Oscar. But, since water beliefs and twater beliefs have the same functional roles, the Ramsey sentence of T would abstract away from the difference, and could be used to generate ascriptions of the same functional states to the two subjects.

But the proposal does not apply to Alf and his "arthritis" concept. Recall that Alf, prior to his enlightenment, has a different concept from ours and the doctor's. This was shown by the argument of the previous chapter. Alf might acquire the prestige "arthritis" concept while retaining his original one. We could tell that the two would be different because they would function differently in his psychology. The idea behind functionalism is that the difference between the two concepts is one of functional role, and functional role is all there is to cognitive content. The problem is that we do not appear to have a psychological theory that tells us about Alf's original "arthritis" concept and how it differs from the prestige concept.

Whether one sympathizes with social externalism or not, it is agreed on all sides that it is natural in common-sense psychology to use the term "arthritis" to talk about Alf's original concept. I argued that, even so, Alf does not have the concept of arthritis. When we use that term in that way, we use it to express a different concept. So what the internalist needs to do, is to say something about what this concept is, how we might talk about it, and so on.

DRNF$_1$ alone simply does not address such questions. Rather, it would enter the story only after we had found a way to characterize Alf's concept. So, in the end, DRNF$_1$ does not manage to finesse the issue of what an unreduced narrow psychology would be like.

The functionalist might attempt to address this problem by developing a theory that specifies the functional role of Alf's "arthritis" concept, or the states in which it features, without relying on a previously available content ascription. DRNF$_2$ develops this possibility.

DRNF$_2$: conceptual role semantics

Suppose that cognition involves the construction and manipulation of mental representations. Mental representations are like words and sentences of natural language, in that they have physical and syntactic properties, as well as semantic ones. The physical properties of natural language sentences vary across different particular tokens or occurrences. Witness "tigers growl," "TIGERS GROWL," and a spoken occurrence of the same sentence. But all of the different occurrences have the same syntactic and semantic features. They all consist in the noun, "tigers," followed by the verb "growl." And they all mean that tigers growl. We can think of mental representations in just the same way. They would presumably be physically realized by patterns of connectivity among neurons, or by patterns of cell firing. At a more abstract level, these representations have syntactic properties. And these properties can be used to classify types of representation in a subject's repertoire, just as we classify various sounds and shapes as tokens of the sentence type "Tigers growl."

We can think of cognitive states in terms of relations to mental representations. For example, to believe that it will be sunny tomorrow is to have a representation which means that it will be sunny tomorrow, playing a particular role in your head. To desire that it will be sunny tomorrow would be to have the same representation playing a different role in your head. When you think and reason, cognitive systems in your brain construct and manipulate representations. Suppose, for example, that you believe that it will be sunny tomorrow and you believe that if it will be sunny tomorrow, then you should water the lawn, and you infer from this that you should water the lawn tomorrow. What happens is that you start out with two representations playing the belief role: *It will be sunny tomorrow* and *If it will be sunny tomorrow, I should mow the lawn*. Your cognitive systems react to the presence of these two representations by constructing a third one: *I should mow the lawn tomorrow.*[9]

The conceptual role of a representation is the role it plays in cognition, where this is construed in terms of relations among representations. The representations are syntactically defined, and the relations are specified without reference to their contents. The relations might involve certain patterns of association among representations: e.g., *water* associates with *transparent*, *potable* and so on. Or, more simply, they might be causal relations: a representation's conceptual role would then be determined by its tendencies to interact causally with other representations, given the way it is treated by the cognitive systems that act on it. $DRNF_2$ is the thesis that we can identify the narrow content of a mental representation with its conceptual role, so understood.[10]

$DRNF_2$ gives the functionalist something to say about Alf's "arthritis" concept. Alf has a mental word, *arthritis*, corresponding to the term "arthritis" in his idiolect. And, like other mental representations, this one has its own particular conceptual role, which is its narrow content. On this view, we don't need any prior content-ascribing psychology, any unreduced psychological theory, that can provide the content of the concept. Rather, the theory that gives its narrow content just is the theory that gives its conceptual role.

The problem with $DRNF_2$ is that the kind of difficulties discussed on pp. 92–95 become exacerbated. The conceptual role of a representation depends on structural features of the cognitive system to which it belongs, such as memory buffer space. And such matters do not seem relevant to cognitive contents. Suppose, for example, that Lena and Zena appear to have the same "water" concept. They share attitudes to "Water is H_2O; it is transparent, odorless and tasteless; it boils at 212 degrees F," they would agree on which presented samples to call "water" and which not, and so on. However, Lena has drunk too much alcohol in her life, and this has reduced her ability to keep track of complicated inferences, while Zena has remained moderate and sharp. This results in substantial differences of conceptual roles across all their representations. Whenever any representation features in a complex inference, the subjects will arrive at different results. Prima facie, it appears that Lena's and Zena's "water" concepts have the same content. But the concepts do not have particularly similar conceptual roles in the two subjects.

The example indicates that similarity of conceptual roles of a purely abstract kind, a kind that draws only on the pattern of conceptual relations that a representation enters into, does not seem to be a good diagnostic of content. Perhaps functionalists are right to think that associations among concepts can be a good guide to content. Perhaps it is right, for example, that if two people associate *drinkable, transparent, liquid at normal temperatures, boils at 212 degrees F* with a concept *c*, then *c* is a *water* concept, or something like one. But it is far less plausible that someone's capacity to keep track of complex inferences is very relevant to ascertaining which concepts they have. The trouble with $DRNF_2$ is that it doesn't offer a way of distinguishing these different aspects of conceptual roles. The skeptic will conclude that the reason for this is that the diagnostically important aspects of conceptual role depend on the contents of the concepts involved. But then it is content that is doing the work, not conceptual role.

While $DRNF_1$ and $DRNF_2$ both have difficulties with providing contents that would be shared by subjects with structurally different cognitive systems, $DRNF_2$ is substantially worse off in this respect. This is because it has fewer resources to draw upon. $DRNF_1$ begins by accepting some kind of content-using psychological theory. It could, for example, begin by accepting a theory that attributes the same "water" concepts to Lena and Zena. The hope would be that the Ramsey sentence of this theory could be used to generate ascriptions of the same functional state to both subjects. The worry would be that the project might fail, because the theory doesn't specify the functional role of a concept in all possible types of system.

$DRNF_2$ has a completely different starting point. It supposes that we could identify mental representations in purely syntactic terms and that we could develop a theory of the conceptual roles of these representations without any reference to their cognitive contents. It cannot draw on a previously accepted content-ascribing psychological theory. Rather, it has to rely on conceptual roles specified by a theory that only has access to the syntactic properties of mental representations. And it does not seem particularly likely that such a theory could provide a good basis to define a workable notion of similarity of conceptual roles across cognitively dissimilar subjects, such as Lena and Zena. It is, of course, possible that some suitable notion of similarity of conceptual role could be found. It might, for example, turn out that some mathematical analysis of the conceptual roles could filter out the similarities we want. But the onus is on the proponent of $DRNF_2$ to provide a positive reason for thinking that this will indeed be the case.

So let us move on to less reductive accounts of narrow content.

4.2 CHARACTER

Character is best understood in the context of demonstrative concepts. We begin with a discussion of these.

Demonstrative concepts

Demonstrative concepts are those typically expressed by demonstrative pronouns such as "I," "she," "this," and "that skinny bloke" in deictic use (i.e. when they are not anaphoric on other expressions). The main line of argument from chapter one extends directly to demonstrative concepts, showing that they have an extension-independent content. An example will illustrate.

$Fred_1$ is in the jungle. There are rustlings in the undergrowth, and $Fred_1$ has the impression of fleeting movement. He suffers a frisson: "that snake could be dangerous." And he is right, since it is a cobra, Charlie. Much the same happens to $Fred_2$ and $Fred_3$ on TE_2 and TE_3. But on TE_2 it's a young python, Peter. And on TE_3 there's no animal there at all, just the wind and $Fred_3$'s imagination. We should hold that all three Freds have a concept expressed by "that snake" and that all three concepts share a cognitive content.

The argument proceeds as before. We begin with the empty case, TE_3. We argue that when $Fred_3$ says "that snake could be dangerous" he is expressing a genuine belief with real cognitive content. We then argue that a good psychological theory would attribute the same belief with the same cognitive content to the other twins.[11]

As before, we conclude that the twins' have concepts that share a cognitive content and that differ at most in their extension. However, it is clear in these cases that the concepts also differ at least in extension. $Fred_1$ has a concept that picks out a specific cobra, Charlie, and his thought is true iff Charlie might be dangerous. $Fred_2$ has concept that picks out Peter. His thought is true iff Peter could be dangerous. And $Fred_3$ has an empty concept, one that does not refer to anything.

Since the extensions of the target concepts differ in the ways specified, we cannot account for the shared cognitive content in terms of them. Rather we need some kind of narrow content, or mode of presentation that can present different objects, or fail to present any object at all. I think there are only two possibilities.

One suggestion might be to account for narrow content in terms of extension conditions. We might, for example, attribute to the Freds' thoughts *that snake could be dangerous* a content like: *there is a unique snake, x, causing this visual experience, and x is dangerous.* This is John Searle's view (1983). Here the only difference in the extension conditions of the twins' thoughts is that they make reference to different visual experiences: $Fred_1$'s thought refers to Fred's visual experiences, while $Fred_2$'s thought makes reference to $Fred_2$'s visual experiences. Since the visual experiences are within the subjects, the extension conditions are themselves intrinsic to the subject.[12]

We might want to recognize a variety of intentional content along those lines. But it is not plausible as an account of cognitive content. Searle himself explicitly denies that the thoughts themselves have these conditions as their representational contents. In thinking about the snake, Fred$_1$ doesn't think about or represent his own visual experience. Nor, indeed, need he have the capacity to do so. Rather, we theorists write down those conditions as part of an account of how the thoughts latch onto the world. Given this, we cannot identify the suggested extension-conditions with cognitive contents. After all, *that snake could be dangerous* and *there is a unique snake, x, causing this visual experience and x could be dangerous* presumably have the same extension conditions under the current proposal. But they differ in their cognitive contents and psychological roles.

The remaining possibility, due to Burge (1977), is that we think of the cognitive contents of demonstratives as extensionally (i.e. truth-conditionally) incomplete. The truth conditional content of demonstrative thoughts is composed of two factors. One factor is just the referent (if any) of the demonstrative concept. The other factor, which provides the cognitive content, has the logical form of an open sentence. To illustrate the basic idea: suppose that Peggy is observing the Chrysler Building and thinks to herself: *that's great*. The demonstrative concept, *that*, works like a free variable in formal logic. The cognitive content has the form "x is great". The sentence "x is great" is not truth-evaluable by itself. Rather, it is true or false relative to assignments of values to the variable: "x is great" is true relative the assignment of object α to "x" iff α is great. When Peggy thinks *that's great*, her demonstrative concept refers to the Chrysler Building. We now get truth conditions for the thought in the context: it is true iff the Chrysler Building is great. The truth conditions of the thought are composed of the incomplete cognitive content and the Chrysler Building itself.

Cognitive contents of demonstrative thoughts are object-independent. They can therefore pick out different objects. Twin Peggy thinks the same cognitive content of the Twin Chrysler Building. But the thoughts have different objects, hence different truth conditions. And they can be entertained in the absence of any appropriate object, when the demonstrative fails to refer.

Demonstratives and character

Pure demonstratives (*this, that*), indexicals (*I, now, today, yesterday*, etc.) and complex demonstratives (*that tall building, this small cat*) have neither context-independent extension nor context-independent extension-conditions. But they do have context-independent characters. Character consists in context-relative extension conditions: functions from contexts of thought to extensions.[13] One can think of contexts as consisting in, or including, a sequence of relevant items: thinker, time, place and so on. The character of *I* maps any context c, onto the thinker in c. The character of *here* maps any context c, onto the place of c. The character of *today* maps any context c onto the day encompassing the time of c. And so on.

Kaplan (1990a), following Perry (1977), suggests that the cognitive content of a thought or concept goes with its character. As Perry points out, if Paul thinks *I am about to be attacked by a bear* and Peter thinks of Paul *he is about to be attacked by a bear*, Peter and Paul will behave differently: Paul rolls up in a ball and Peter goes to get the park warden. Extension therefore does not determine psychological role. But if Peter and Paul both think *I am about to be attacked by a bear*, then they will do the same thing. This time Peter rolls up as well. Can we, then, identify cognitive content with character?

Definitely not. If characters are functions from contexts of thought to extensions, then thoughts with the same character can have different psychological roles.

Compare: *yesterday*; *the day that would be referred to by a present utterance of "hier" in French*; *the day before the day of this thinking*; *yesterday, or, if today is Sunday, the most recent Saturday*. All of these map the context of thinking onto the day prior to the day of thinking. But obviously they have different cognitive contents.[14]

What, then, is the relation between character and cognitive content? Some aspects of character seem to be relevant to cognitive role. For example, it is hard to see how a subject could think of something as *now* and not represent the something as a time, or think of something as *here* and not represent it as a place. Perhaps representations of these aspects of character are actually included in the demonstrative concepts. Or perhaps they are just closely associated with them. But either way, they constrain cognitive role and it seems reasonable to count them as determinants of the cognitive content of a concept.

But not all aspects of a demonstrative concept's character will be represented by the thinker or her cognitive systems. Characters are functions from contexts to extensions. So a representation of character ought to quantify over contexts. However, there is no reason to suppose that demonstrative thoughts necessarily quantify over contexts, nor even that quantifications over contexts have to play a role in cognition. Quantifications over contexts are part of a theory of how demonstrative concepts work. But one doesn't need the theory in order to have the concepts.

It does not even seem to be the case that the distinguishing aspects of a demonstrative's character, the ones that distinguish one type of demonstrative from another, need to be represented in cognition. For example, the character of *I*, as noted, is something like "the thinker of *c*" (as opposed to "the time of *c*" and so on). It is unlikely that every cognitive system able to entertain *I* thoughts is able to represent thinkers as such. People with autism, for example, appear to lack the capacity to form these second-order representations (Leslie 1987, 1994; Baron-Cohen 1995). However they do not lack the capacity to think of themselves. They merely fail to represent themselves as thinkers. In any event, even if there is some connection between *I* concepts and representations of thinkers as such, it really does not seem that when one thinks of oneself with one's *I* concept, one must thereby represent oneself as a thinker.

Interim conclusion

The cognitive content of demonstrative thoughts can be accounted for largely in terms of general concepts. These thoughts include a purely referential component that functions like a free variable, along with general, predicative concepts. The demonstrative concept may perhaps be complex and contain predicative material within it. This is clear in the case of explicitly complex demonstratives, like *that skinny cat*. Predicative material is also included in cases where aspects of character are represented by the subject or her cognitive systems. Perhaps there is quite a lot of predicative material present in typical demonstrative concepts that we do not tend notice. But there is not enough to provide complete truth-conditions for typical demonstrative thoughts (see Burge 1977, 342–343). The general form of these thoughts is either $[x][F]$, where the demonstrative concept is just the purely referential free variable, or $[Gx][F]$, where the demonstrative component combines a predicative concept and a free variable.

The referential component, x, has very little cognitive content of its own. Its cognitive content is simply that of *that*, or perhaps even less, since *that* differs from *this* in that the former tends to pick out distal objects and the latter, proximal ones. The burden of cognitive content of demonstrative thoughts is therefore placed with the associated general concepts.

The next subsection discusses whether materials from the account of demonstratives can help us understand the cognitive content of general concepts. One thing is clear from the outset, however. The account cannot extend to all general concepts. For if it did, then the cognitive content of all concepts would be the same, the exiguous content of the free variable.

Character and general concepts

As Burge (1982) and others have pointed out, terms like "water" and "arthritis" are not like normal indexicals. Their extensions appear to be context-independent rather than context-dependent. As Oscar moves from place to place, persists through time, addresses different people and so on, occurrences of concepts *here*, *now*, *tomorrow*, *you* and so on pick out different extensions. But, according to externalist intuitions, if he thinks *water* or *arthritis* in any context, he always picks out the same extension: water or arthritis respectively.

However, as Fodor (1987, chap. 2) points out, if we individuate contexts of thought four-dimensionally, then we leave open the possibility of extending the indexical account to general concepts.[15] If Oscar had gone to Twin Earth and stayed there for long enough, his term "water" would have ended up picking out XYZ rather than H_2O. Perhaps we could think of his associated concept in the same way: he retains the concept, but it alters its extension. We might then think of contexts as including, in addition to times, places, etc., what we can call "home environments." Your home environment is the environment in which your concepts have become embedded, the environment that contains the kinds that stand in extension-determining relations to your kind concepts and the experts who warrant your deferential allegiance.

If Oscar had been in Twin Oscar's home environment (had lived on Twin Earth for a suitable period) then his "water" concept would have had the same extension as Twin Oscar's. If Twin Oscar had been in Oscar's home environment then his concept would have had the same extension as Oscar's. And, indeed, Oscar and Twin Oscar's "water" concepts would pick out the same extensions in any home environment. In this sense, their concepts have the same character.

The proposal is that we might think of narrow contents in terms of this extended version of character. Fodor's own presentation of the proposal suffered from problems of underspecification. He offered no adequate answers to questions like the following: "Which function from context to extension is the character of Oscar's 'water' concept?" "What kind of vocabulary do we use to characterize these functions and how would this vocabulary feature in a psychological theory?" Can we do better? Let us consider natural kind concepts first, then move on to other deferential ones.

As with demonstratives, it is clear that the character of a natural kind concept does not determine its cognitive content. The character of Oscar's "water" concept could perhaps be expressed by "the substance in my home environment that is colorless, odorless, tasteless, and takes liquid form at normal temperatures." If so, it could also be expressed by "the substance in my home environment that is colorless, odorless, tasteless and takes liquid form at normal temperatures and which might or might not be H_2O." The concepts associated with these lengthy descriptions have the same character. But they have different cognitive contents.

If the discussion of demonstratives was on the right track, then the account of cognitive content cannot draw directly on character, but can draw on aspects of character that are represented by the subject or her cognitive systems and that constrain cognitive role. One might, then, hope to give an account of narrow content with these materials. But the optimism would be misguided.

It is not particularly easy to think of plausible descriptions of the character of kind concepts. What would be the character of *tiger*? Perhaps "yellow-brown, black-striped, carnivorous, maneless feline in my home environment." But this will not do. When applied to Earth, or at least nearby possible Earths, it fails correctly to fix the extension of *tiger*. There are, or could be, albino tigers, stripeless tigers and maned tigers. Stereotypical features are not essential ones. A more promising attempt is: "that kind in my home environment, typical members of which are yellow-brown, black-striped, carnivorous, maneless and feline." There are two problems with this proposal. The first concerns the more superficial, observable features mentioned in the description, the second concerns the deeper ones: carnivorous and feline.

Notice that the choice of superficial features (borrowed from the *Concise Oxford English Dictionary*) is somewhat arbitrary. Why should "maneless" be included, but not "furry," "fierce," "apt to growl and roar"? What decides such questions? The answer seems to be that a feature should be included only if we could not be wrong in supposing that typical tigers possess it. If we could, then the description does not provide any direct constraint on the extension of the concept in a home environment. But then none of the superficial features should be included. If we stretch our imagination, we can imagine finding out that no tigers are yellow-brown (it was a strange trick of the light) or black-striped (they were shadows), or maneless (we failed to notice the short, almost invisible, whiskery manes).

We are left, then, with "carnivorous" and "feline." But these are clearly insufficient to account for the cognitive content of *tiger*, since *tiger* differs from *leopard*, *domestic cat*, and the rest.

The kind terms "feline" and "carnivorous" also raise particular problems. Should they be included or not? On Robot Earth, there have never been any real tigers, but there are very convincing robot imitations. These are neither feline, nor carnivorous. The humans there are our twins. The issue is whether these twins of ours refer to robot tigers when they say "tiger." If they do, then the terms should not be included. (Suppose that "feline" means *feline* on Robot Earth: there are real cats, lions, etc., there). If they do not, then the terms should be included. If they do, then they should not.

One way to get a handle on the question is to consider whether, if we found out that this Earth was Robot Earth, we should say "there aren't any tigers, only robot fakes" or "tigers are robots." Intuitions divide on this question. But either proposal leads to trouble. If they are not part of the character then we have no description of the character left at all! But if they are part of the character, then the character has yet to be shown to be narrow, for it is described in kind terms that, according to the externalist, have wide contents.

To be acceptable to the internalist, then, "carnivorous" and "feline" must be susceptible to the same type of analysis that we proposed for "tiger." But evidently the same problems will just recur. The observable features, "furry" and so on, are epistemically defeasible, hence unacceptable. The deeper features, "mammal," "quadruped," and so on, are natural kind features. In the final analysis we are only allowed to draw on relatively superficial features, which are not actually determinants of character.

In sum, it appears that there is no serious chance of accounting for narrow content of natural kind concepts in terms of descriptive material that contributes to determining character. This result should come as no surprise. For the motivation for the account was in part externalist. The point was to uphold externalist claims about extension conditions and provide an additional account of narrow content. But the externalist view of natural kind concepts is importantly similar to the externalist view of proper names. A major component of what fixes the reference of most proper names, on this view, consists in causal relations between users of the name and the referent. We learned from Kripke (1980) that those relations largely screen out the relevance of associated descriptions. Associated descriptions usually play little or no role in fixing reference. All that we have found is that the same applies to kind concepts, on the externalist view. Associated descriptions do not contribute much to fixing reference, even within a context. A two factor theory designed to satisfy externalist views of extension, while accounting for narrow cognitive content in terms of character-fixing descriptions, was always going to be unworkable compromize.

One might attempt to evade the difficulties by introducing meta-linguistic material into the descriptions of character. The character of *water* might be or include *stuff called "water" in my home environment*. This is also the natural account of the character of deferential concepts. On the externalist view, an important factor in fixing the extensions of Alf's and Twin Alf's "arthritis" concepts is their deployment of the word form "arthritis" and their deference to the experts in their home environments.

But the meta-linguistic account is hopeless. There are at least two problems. The first is that all, or nearly all concepts are deferential (see Burge 1979, 80–83). Narrow content then ends up being almost exclusively about words. Consider "arthritis." Is its character merely "thing called 'arthritis' in my home environment"? If so, then this clearly does not suffice to determine cognitive content. Consider a cousin Earth in which "arthritis" means *water*. Then, on this proposal, the cognitive content of a normal Earth individual's concept of arthritis is the same as that of a normal cousin Earth individual's concept of water. The cognitive content of *arthritis* is far richer than what is provided by that meager description of its character. So we must enrich the description. We might try, for example, "disease called 'arthritis' in my home environment." But the problem is that *disease* is also a deferential concept. So, for that matter, is the concept of one's home environment. So we are left with "thing called 'disease' called 'arthritis' in my thing called 'home environment'." But now we are back to the original problem. Merely adding more meta-linguistic material does not enrich the contents in the right way.[16] Metalinguistic material, no matter how extended, does not determine cognitive content.

The second problem is that cognitive contents don't determine metalinguistic characters. French and English near twins could presumably share a concept of arthritis. But the meta-linguistic characters would be different, the French one including "called 'arthrite' " where the English one has "called 'arthritis'."

4.3 DESCRIPTIONS

A related idea would be that narrow content might be accounted for directly in terms of associated descriptions, whether or not these determine reference in context. I do, after all, think of tigers as typically being yellow-brown, black-striped, carnivorous, maneless and feline. So perhaps the cognitive content of my *tiger* concept is given by these descriptive concepts.

This proposal suffers from some familiar problems. First, we need to decide which of the various descriptions I associate with tigers is to be counted as relevant. How might we select among, for example, "yellow-brown," "black-striped," "carnivorous," "maneless," "feline," "animate," "furry," "fierce," "apt to growl and roar" and "indigenous to India"? The more superficial descriptions seem more or less on a par. None has a privileged role in determining cognitive content. The kind terms have a greater influence on cognitive role than the superficial ones, since they are embedded in a biological theory and are rich in inductive consequences. The fact that tigers are animals generates more interesting consequences than the fact that they are yellow-brown. Moreover, it is reasonable to regard animacy as a definitive property of tigers. That is why robot tigers are ipso facto not tigers but albino tigers are.

As we have already seen, the associated kind concepts do not suffice to determine cognitive role, because we know of too many species of carnivorous felines. So the superficial features must be included somehow. But there really seems to be no way to include some and exclude others. So we must include them all. The natural proposal is that the cognitive content of a concept is given by what we might call its associated theory: an integrated network of associated descriptions. Will this proposal fare better than the previous ones?

The idea that the content of natural kind concepts are fixed by role in theories is popular among psychologists (e.g., Carey 1985, Keil 1989, Spelke 1988; for criticisms see Fodor 1997, 112–119). One might think of such roles in functionalist terms. Select the theory, then Ramsify it to get a specification of the theoretical roles of all the target concepts in terms of their relations to each other and, perhaps, to some privileged set of concepts—maybe observational or non-theoretical ones—that are left undefined.

In the present context the idea would be that we could distinguish the kind terms that are problematically wide from other terms that are not susceptible to externalist Twin Earth experiments. The problematic kind terms would then be accounted for purely in terms of abstract relations to each other and to the internalistically acceptable residue. Thus we might take a biological theory, T, and distinguish the theoretical terms from the others. The former might include, e.g., "tiger," "panther," "feline," "carnivorous," "mammal," "animal" and "species." The latter long list might include "yellow-brown," "striped," "apt to growl," "fierce," "edible," "dangerous" and "large."[17]

This account becomes extremely holistic. For theories are not self-contained. It is part of my theory of animals that they are physical objects. But of course *physical object* is a core concept of another large theory. I also believe that some animals are cognitive systems. So, to get the complete story, we would have to Ramsify my biological, physical and psychological theories at the same time. Indeed, given the amount of overlap among theories, it is reasonable to conclude that to Ramsify one, we would have to Ramsify all. Thus to account for a single concept we have to specify the subject's entire theory of the world.

This proposal should appear familiar. It is just a variant of functionalism, rather like $DRNF_1$. The difference is that under $DRNF_1$, the theory to be Ramsified was psychology, the commonsense or scientific theory of the psychological role of concepts. On the present proposal, it is the subject's own theory, the one in which the concepts are embedded, that is to be Ramsified. Nevertheless, it comes to much the same thing. The idea, once again, is to account for the cognitive content of one concept mainly in terms of its relations to other concepts. And it suffers from the same defect. For, like $DRNF_1$, the proposal fails to include an adequate account of the vocabulary a psychologist might use to specify the cognitive contents.

The point, recall, is to describe narrow properties of concepts that we can use in psychological generalizations. It is not enough, for example, just to specify something in common among the twin Oscars' "water" concepts. We need to find shared properties that can be specified in a psychological theory and used to form generalizations subsuming all the twins.

If we could find a principled way of distinguishing some specifiable set of descriptions that gives the cognitive content of a concept, then there would be no difficulty in explaining how cognitive contents function in psychology. We would simply replace references to the concepts with references to their descriptive paraphrase concepts and proceed as normal. If Oscar believes that water is good for plants and Twin Oscar believes that twater is good for plants, then we might say something like: both Oscars believe that the colorless, odorless, tasteless substance in their home environments that takes liquid form at normal temperatures is good for plants. And we could then proceed to provide psychological explanations and predictions in terms of this attribution.

But the functionalist version of the descriptive proposal is not so easy to incorporate in psychology. We can hardly be expected to specify the subject's whole theory of the world every time we wish to attribute a psychological state.

Conclusion

The two factor theories we have considered are not promising. What this suggests is that we should try something else. Rather than trying to come up with an account of narrow content and then figure out how it can do the explanatory work it is supposed to do, I suggest that it would be more fruitful to begin with content as it already appears in psychology and see whether we can treat it as narrow. Chapter 5 pursues the suggestion.

Narrow Content and Psychology

Cognitive content is part of the subject matter of psychology. It is what gives psychology its explanatory and predictive power. Concepts should therefore be taxonomized according to cognitive content. The vocabulary of psychology should subsume concepts with shared cognitive contents under the same terms, and concepts with different cognitive contents under different ones.

The two factor theories discussed in chapter 4 are not well placed to meet this requirement. We do not have an adequate vocabulary with which to describe the narrow factor and frame the psychological generalizations we want. The basic point is simple: once one adopts the externalist view of extensions, it becomes difficult to use normal propositional attitude attributions to state the desired generalizations. On that view, Oscar believes that water is good for plants. Twin Oscar does not. We cannot use the ordinary content clause ("water is good for plants") to characterize the twins.

On certain versions of character or descriptive theories of the narrow factor, a relatively conservative revision of externalist descriptions of twins would be possible: all the Oscars believe that the colorless, odorless, tasteless substance that takes liquid form at normal temperatures in their home environments is thirst quenching. But no appropriate descriptive paraphrases are to be found. Other two factor theories essentially give up on the idea that narrow psychology would look anything like extant commonsense or scientific psychology. Rather, we would have to switch vocabularies altogether to talk about something else, such as functional roles.

I think that both externalism, in many of its manifestations, and two factor theories are based on a common mistake. The mistake is to attach too much significance to the externalist intuitions generated by the kind of thought experiments that Putnam and Burge have given us. There is, indeed, a tendency to think that Oscar believes that water is good for plants, while Twin Oscar does not and that Alf believes that he has arthritis in his thigh, while Twin Alf does not. The mistake is to take this tendency too seriously and then to draw from it a general conclusion about the extensions of concepts: either cognitive content is at least sometimes wide, or cognitive content does not determine extension. Generalizing from such twin cases, then, the conclusion is that psychology (at least ordinary common-sense psychology) is either externalist or in need of radical revision. The only move left for the internalist is then to try to develop and defend some revisionist picture of psychology.

The view I recommend is quite different. I think that psychology as it is practiced by the folk and by the scientists, is already, at root, internalist. The externalist intuitions generated by the focal Twin Earth experiments are simply misleading. They reveal only an accidental and adventitious strand of our psychological thinking. The basic apparatus of psychology does not mandate externalism. So ascriptions of content that are made when practicing good, correct psychology are already internalist: the contents they attribute are already narrow.

On this view, the narrow content of my belief that tigers can be playful is simply this: *tigers can be playful.*

The narrow content of my belief that diamonds are overpriced is this: *diamonds are overpriced*. And so on. To say that these contents are narrow is to say that they are intrinsic to me, hence that any twin of mine would have beliefs with the same contents. Internalism, therefore, does not need to posit any fancy additional notion of content beyond that which is already evidently at work in psychology, that which is already attributed by content sentences of propositional attitude reports. No deep or sweeping revision of psychology is required.

However, if we are correctly, precisely and explicitly to describe Oscar's and Alf's minds, then we do need to revise certain of the particular ascriptions offered by externalists. In Oscar's case the ascriptions are wrong: he does not have the concept *water*, so he has no *water* beliefs. And I think it is simply wrong for us to use the word "water" in the content sentences of our discourse on his thoughts. In Alf's case the ascriptions are perhaps not wrong. But they are misleading. Alf does not have the concept *arthritis*. As I argued in chapter two, if we can say truly that Alf believes he has arthritis in his thigh, then we are using "arthritis" in Alf's way and not attributing the thought that he has arthritis in his thigh.

So, if we are correctly, precisely and explicitly to describe these particular subjects' minds, then we would do well to adopt neologisms. We can say that the Oscars believe that "dwater" is good for plants, and that the Alfs believe they have "tharthritis" in their thighs. The reason we need to adopt neologisms in these cases is not that the contents we attribute are narrow. It is simply that our words "water" and "arthritis" do not express Oscar's and Alf's concepts. So if we are to use straightforward propositional attributions correctly, precisely and explicitly, we need some new words.

Do we really understand "dwater" and "tharthritis"? What is in the extensions these concepts? How are we to find out? These are good questions. I will offer a brief and preliminary explanation of the particular terms "dwater" and "tharthritis" now. The remainder of the chapter will be devoted to general issues about contents and how we can find out about them.

I think we probably understand "dwater" rather well. It means roughly "waterlike substance" or "hydroid stuff." "Tharthritis" *per se* is not really an issue, since there are many ways of expanding Alf's biography that would endow him with different concepts. Perhaps "tharthritis" means roughly the same as "rheumatism." Or perhaps Alf has a more detailed misconception. He might, for example, think that tharthritis is a specific kind of hereditary auto-immune disease. The details are not so important.

"Dwater" includes both XYZ and H_2O in its extension. It includes very unhydroid things like oil, alcohol, sugar, electricity and oxygen in its anti-extension. And there are probably many substances that are in neither, such as heavy water (D_2O) and other substances that resemble normal water in some ways but not others. "Dwater" is neither true, nor false of these. The concept Oscar expressed by the term "water" is therefore substantially different from the concept I, and probably you, dear reader, express by that term. Unlike Oscar, my "water" concept is crucially conditioned by my belief that water is H_2O: I think that it is a deep truth that if anything is not H_2O, then it is not water. If I understood metaphysical necessity, I would probably think that it was a metaphysically necessary truth. Mine is a genuine natural kind concept, one that applies to an extension defined in terms of deep structural properties, as discovered by science. Oscar's is not. It is a motley concept that applies in virtue of relatively superficial features.

The extension of "tharthritis" depends upon how the character of Alf is fleshed out. If he had the articulated misconception mentioned above, then a decent first hypothesis about the extension would be that it includes any hereditary auto-immune diseases that can cause at least some of the kinds of symptoms that Alf has in his joints and thigh. It would exclude, for example, influenza, broken bones and sunstroke. And, like "dwater," "tharthritis" probably generates an area of indeterminacy. This might include, say, non-hereditary auto-immune diseases that produce the same symptomatology as paradigm cases. We would find out more about the details of Oscar's and Alf's concepts just by doing psychology. We could make a start by interviewing them.

I am not going to offer an algorithm for discovering the extension conditions of a concept. I know of no such algorithm. Nor am I going to offer any philosophical theory of content. As I said in the introduction, attempts to formulate such theories seem to me to be premature. Cognitive content is the subject matter of psychology. So one good way to discover the cognitive contents of people's concepts is to do correct psychology. And, although there is no known algorithm for doing psychology correctly, there are certainly some substantial heuristics. In the final section of this chapter I will say something about these, and about why good psychology need not be externalist.

Prior to that, I want to adjust the dialectical balance. For it may seem that the externalist has an advantage, offering more satisfactory answers to questions about the extensions of concepts than I can offer. It can seem that Putnam and Burge have offered principled and complete answers to questions about the extensions of Oscar's and Alf's concepts, answers, moreover, that do not leave inelegant areas of indeterminacy. This is all an illusion.

The considerations that Putnam and Burge offer do not lead to algorithms for finding out what is in the extension of a concept. Nor could they lead to satisfactory theories of content, if what I have argued in chapters 2 and 3 is along the right lines. I will argue that in general we are all in the same boat (or rather similar boats, anyway) when it comes to addressing questions like "What is x's concept of Y?" and "How do we find out?"

In more detail, the plan is as follows. I will begin with some criticism of Putnam's account of the role of natural kinds in determining extension conditions. This will help restore the dialectical balance and will provide some motivation for the claim that Oscar's and Twin Oscar's dwater concepts are true of both XYZ and H_2O. I then turn to the experts, and argue that it cannot be the case that experts in general play the role that Burge outlines. The point of this particular argument is not that Burge's view is false. He does not claim that the account applies across the board. Rather, the point is that it fails to apply to many concepts. It follows that we are all in similar boats at least some of the time. Externalists and internalists alike will need to say something about the extension conditions of many concepts without appealing to experts. This conclusion will motivate the claim that many of the difficulties for an internalist account of concepts like Alf's "arthritis" are actually difficulties for everybody. The concluding section defends the use of neologisms and sketches an internalist picture of psychology.

5.1   AN INTERNALIST ON TWIN EARTH

Problems for Putnam's model

Interestingly, Putnam (1975a) did not really give an argument for the conclusion that "water" in 1750 had the same extension as it does now, referring only to H_2O.

Rather he assumed it, perhaps on the basis of intuition. He then gave a sketchy model of how extensions of natural kind terms are fixed. It is a sort of "just so" story with an air of plausibility to it. Humans encounter some number of samples of the kind. They use a word, W, to refer to the samples. And they intend the word to generalize beyond encountered samples to future and counterfactual cases, roughly as follows: "W is to be true of anything that has the same underlying nature as these instances, and of nothing else." In the case of water, the underlying nature is of the sort appropriate to liquids. Thus, "water" applies to this stuff (pointing to some samples) and to all and only other samples of the same liquid. What makes something the same liquid as these samples, what fixes the appropriate underlying principle of classification, is to be decided by science. Science is in the business of carving nature at its joints, and so, when it succeeds, finds the correct instantiation for "same liquid."[1]

This is an initially appealing story, and it does fit with some intuitions as well as with some aspects of scientific practice. However, Putnam's model cannot be quite right. A second example will show why. It is adapted from LaPorte (1996).

In the 1920s, a group of Earth scientists are sent to a twin Earth. They find that a certain kind of liquid is common there. It looks rather like typical Earth water. Interested, the scientists wonder whether what they have found is water. On examination, they find that while it resembles typical Earth water in many respects, there are some important differences. For example, it boils and freezes at slightly different temperatures than does normal H_2O. And, importantly, it is not conducive to terrestrial life, but rather is fatal to plants and animals from Earth.

The scientists go on to examine its internal constitution. They find that this, too, is interestingly different from typical water. Rather than being composed of normal hydrogen and oxygen, it is composed of oxygen plus a hitherto unknown component, which, like, normal hydrogen, has only one proton and one electron, but has an additional neutron. They decide that they have found a new element, which they call "deuterium" and give the hydroid substance the chemical description "D_2O." Having discovered that this stuff is both macroscopically and chemically different from water, and being somewhat Putnamian in their views about kind terms, they feel they have established that it is definitely not water. They call it "deutroxide."

When they return to Earth they report on their discovery, emphasizing how they had established that deutroxide is not water. Earth scientists, having already discovered deuterium oxide themselves and regarded it as a variety of water, heavy water, are unimpressed.

LaPorte points out that neither group of scientists is right and neither wrong.[2] Nothing about the nature of the world, nor the initial usage of "water," determines whether D_2O should or should not have been called "water" when it was discovered. We should conclude that "water" was neither true nor false of D_2O prior to 1920. The fact that scientists classified D_2O as a kind of water, rather than not, was presumably due in part to contingencies concerning the distribution of D_2O relative to H_2O (small amounts of D_2O tend to occur in normal water and do not occur separately in large volumes).[3]

We can already see that Putnam's initially appealing account of the extension of natural kind terms cannot be quite right. It is not as if we had initially pointed to some samples of water and said "These samples and all and only samples of the same kind of liquid will be called 'water'" and left it up to science, carving nature at its joints, to tell us what the same kind of liquid really is. Or, to the extent that we do use terms with some such idea in mind, we do not thereby succeed in fixing their extensions.

Mark Wilson (1983) provides other examples of adventitious factors affecting the evolution of natural kind terms. Here is one. "Grant's zebra" was originally applied to a particular species of zebras native to Kenya. A different subspecies of the same species of zebra found in Rhodesia was called "Chapman's zebra." It was discovered that the two subspecies interbred near the Zambezi river. "Grant's zebra" is still used only to refer to the Kenyan subspecies and not to the Rhodesian one. But suppose that Chapman's zebras had not been given a special name and had first been investigated near the Zambezi. Then it would have been natural for the scientists to say: "There are two subspecies of Grant's zebra." The term would then have ended up applying to the whole species. We should conclude that "Grant's zebra" was, originally, neither true nor false of Chapman's zebras.

In Wilson's example, it is simply the order of discovery that determines the developments in usage. But there are many further reasons why Putnam's model is inaccurate. Many terms that might have ended up extending over a natural kind in fact did not. Our taxonomies reflect a wide variety of contingencies that have nothing to do with nature's own joints. The expectations of scientists constitute one obvious factor. And beyond that, there are a whole variety of human interests, the whole plethora of non-scientific activities for which we develop our taxonomies.

Neither "cat" (in its broadest use) nor "whale" picks out a biological kind, even though in both cases one is at hand. Whales are large cetaceans (the order containing whales, dolphins and porpoises).[4] Size matters. But the term could easily have come to cover the other cetaceans as well, just as "water" came to cover D_2O. "Cat," in its broadest use, covers the family Felidae, but includes also members of other families, such as civet cats and polecats.

The great heterogeneity of the development of usage is nicely illustrated by gems.[5] Diamonds have the same chemical composition as coal. But since diamonds and coal are so different in ways that matter to us, we don't call coal "diamonds" nor diamonds "coal." "Sapphire," in its broader use, applies to all gem varieties of corundum (Al_2O_3) except ruby, according to most authorities. The original Greek term probably referred not to sapphires, but to lapis lazuli. "Ruby" applies only to red corundum. "Topaz" applies to aluminum silicate ($Al_2(SiO_4)(OH, F)_2$), no matter what the color. However, the original term, used by Greeks and Romans, applied not to aluminum silicate, but to yellow corundum, i.e. yellow sapphire. Emeralds and aquamarines are both beryl ($Be_3Al_2Si_6O_{18}$), the former green, the latter blue. Jade (as Putnam pointed out) comes in two varieties, jadeite and nephrite. But "jade" almost became a natural kind term. LaPorte found the following entry in an old reference book:

Jade, a name applied to certain ornamental stones ... belonging to at least two distinct species, one termed nephrite and the other jadeite. While the term jade is popularly used in this sense, *it is now usually restricted by mineralogists to nephrite.* (*Encyclopaedia Brittanica*, 1911, 122; quoted in LaPorte 1996; emphasis is LaPorte's)

So, even if most or all of the samples to which a pre-scientific term is applied happen to belong to a natural kind, there are various possibilities for the development of usage over time. First: there will often be more than one natural kind subsuming all the samples ("water," "Grant's zebra"), and nothing makes one a more likely candidate than some others. Maybe there will always be more than one kind available, since natural kinds come in hierarchies: e.g., subspecies, species, genus and so on. Second: the term may well end up applying to an arbitrary subclass of a natural kind ("emerald," "sapphire," "whale"). Third: the term may end up applying to a motley of two or more natural kinds ("jade," "cat"). Last (and perhaps of least significance), it may even end up applying to a kind that the original samples didn't belong to ("topaz").[6]

Putnam's model is evidently far too simple. It just is not the case that by using a term to apply to samples of a natural kind, the term gets to have that kind in its extension. And this shows, at the very least, that the question of what a prescientific term extends over is extremely difficult for a Putnam style externalist to answer. There is no more reason to suppose that all and only varieties of H_2O were in the extension of prescientific "water,"[7] than that all and only cetaceans were in the extension of prescientific "whale," or felidae in the extension of "cat," or beryl in the extension of "emerald," etc.

Indeed it does not seem that there is any good reason to think that a pre-scientific term ever extends over a natural kind. It is true that, as in the case of "water," a term may end up applying to a specific natural kind after pivotal scientific discoveries. But any such term might have ended up either not applying to a natural kind at all, or applying to a different one.

What I propose as a more plausible view than Putnam's is that prescientific terms for natural phenomena apply to motleys. A motley may consist in several natural kinds, or in a collection that includes some, but not all, samples of a plurality of natural kinds. Motley terms are common: "tree," "river," "rock," "metal," "fish," "influenza." When a term becomes a natural kind term with the development of science, its extension alters. Competent users come to regard some particular scientific principle of classification as correct, and so begin to use the term in line with that. When this happens, the extension may enlarge, or shrink or alter its boundaries in both directions. But before the scientific principle is known and more or less explicitly adopted, there is nothing that ties the word to a unique natural kind.

Motleys

The term "water" in 1750 and Oscar's *dwater* concept, then, applied to a motley. More specifically, it was a term the extension conditions of which did not confine it to any specific natural kind, but left open the possibility of it being true of many different kinds. Had there been any XYZ in the universe, the term and the concept would have been true of it. For those still inclined to resist this claim, here are some more Twin Earth experiments that should increase its credibility.

We begin with one adapted from Wilson (1983) and pursued by Larson and Segal (1995). We return to 1750 Earth and Twin Earth. It turns out that a certain fin de (vingtième) siècle hypothesis is not too far from the truth: Earth has long been visited by technologically advanced extra-terrestrials. Rather than being clandestine, these aliens provided a helpful transport service between the two Earths. Earth and Twin Earth people call water and twater "water," unaware that there is any difference between the two. Later, chemistry develops, and chemists notice that what they call "water" comes in two varieties, with different kinds of molecular constitution.

Outlandish as this example is, with its alien taxi service, the account of the development of linguistic usage is plausible. Given that all the subjects involved, scientists and lay persons on both planets, would have been used to calling both water and twater "water," it would have involved a massive and pointless revision of vocabulary to do anything other than continue that practice. This would be so even if Twin Earth had not had its own inhabitants, and there were only Earth people involved. Those Earth people would already have become accustomed to calling twater "water," and would not wish to alter the practice.

The natural conclusion is that "water" on Earth did not refer exclusively to H_2O in 1750. For it appears that it could have come to refer to both water and twater without a change in meaning.

The externalist will of course claim that the word "water" would have undergone a change of meaning at some point during the story. Thus, initially Earth "water" applied only to water. During early encounters with twater, the Earthlings were wrong to call it "water." But after a prolonged period of interactions with twater, the word "water" would come to apply to both kinds of liquid. The externalist description is coherent. But it is not particularly natural or intuitively compelling. On the contrary, the protagonists involved would not have felt there was any change in the meaning of "water" at any point. And that means that if the events had actually occurred, that is not how we would feel about it. We would not feel that the meaning of our word had changed. What drives the externalist view of this case is mainly the Putnamian idea that in 1750 the extension of "water" was already confined to all and only varieties of H_2O. But we have already seen that the idea is not plausible.

It is true that, in the example, the protagonists are ignorant of the difference between XYZ and water, so we should not automatically attach too much significance to their tendency to call XYZ "water." One must consider whether they are making a mistake. But it does not seem that they are. Suppose that a number of the travelers are suspicious of the apparent similarities between the two planets. They might simply abstain on the question of whether twater has the same underlying nature as water and yet still be quite happy to call it "water." They might quite reasonably say, "I think the water around here may turn out to be interestingly different from water back on Earth."

A good way to support this last point is to run a parallel but forward-looking thought experiment. "ME" will serve well as the focal term. Recall that ME is currently identified by its symptoms and that there is no consensus as to their cause or causes. Now, let us imagine that the technology for space travel develops rapidly. We discover a twin Earth. There, too, some subjects suffer from chronic fatigue, exhibiting symptoms similar to those of Earth ME sufferers. On Twin Earth, doctors have not isolated the condition or given it a name. (I tell the story this way to avoid having to deal with the mingling of two different practices of using the term "ME.") Research continues, and it is discovered that ME is caused by two different viruses, one on Earth, the other on Twin Earth. The medical profession continues to use the term "ME" for the condition, but distinguishes the two variants, which, it turns out, have somewhat different patterns of development and require different treatments.

This example begins by considering a term that is applied to what are actually samples belonging to a natural kind: ME on Earth, we are supposing, is an infection by a particular virus. However, at the start of the story, the fact is not known by the experts. The real nature of the disease is undiscovered. At that time, prior to the discovery of the underlying nature of the disease, the extension of the term is not yet confined only to examples of that natural kind. For, when a similar syndrome is discovered on Twin Earth, doctors find it natural to apply the term to that as well. And they would do so—or so it seems to me—even if they were well aware of the possibility that the underlying nature of the Twin Earth condition might be different. Some of them might suspect this, having picked up on some small differences in the symptomatology. However, even they are quite happy to apply the term "ME" when they say "I suspect that there are two different kinds of ME."

So we have a conflict of intuitions. In the original, Putnamian story it seems as though the pivotal scientific discoveries had no effect on the reference of "water." It seems that we now apply the term "water" only to H_2O, and we feel that only H_2O was in the term's extension even in 1750. But in the latter two experiments, intuition goes the other way. In those stories, it looks as though the discovery is crucial. For, prior to the speakers' finding out about the underlying nature of the examples they are talking about, the term they use is not yet a natural kind term. At that point, the term may naturally be applied to Twin Earth counterparts of the terrestrial natural kind. And people might so apply it even if they suspect that the alien samples have a different underlying structure than do the mundane ones. In these experiments, it just does not seem right to insist that the Earth speakers are wrong so to apply the terms. When they call XYZ "water" or Twin ME "ME" they are not making a mistake. And if that is right, then the twin samples must be in the extension of the original Earth terms after all.

You will want to know how we determine what, other than XYZ and H_2O, was in the extension of Oscar's concept of dwater. I will say more about this in the concluding sections. But for now I want to emphasize the dialectical point that this sort of question must be faced by Putnam-inspired externalists as well. Putnam's model of samples and kinds does not by itself offer any satisfactory way of ascertaining the extension of prescientific terms and concepts. For any set of samples, there will be many different candidate extensions grouped by some reasonable principle or other. The challenge for the externalist is then to say something helpful about what determines this range of candidate extensions, and the factors that determine which of the range (if any) is in the extension at any given time. This is the analogue of the challenge facing my proposal.[8]

What is in the extension of "tharthritis"? How could we tell? Burge's account suggests that experts can pin down Alf's apparently rather nebulous personal conception to a clear and determinate extension. I will now argue that experts cannot in general play such an extension-fixing role. For many concepts, when it comes to answering questions of the form "what is in the extension of concept C?," "how are we to tell?" we are all in the same boat.

5.2 EXPERTS AGAIN

Here is a variant on Burge's story. Fred apparently has many ordinary beliefs about pies. He believes that pies have an outer crust, that they can have sweet or savory fillings, that pork pies are cheap and highly calorific, and so on. He begins to learn to cook. One day he buys some pre-mixed pastry, lines a baking tin with it, fills it with an apple and sugar based paste and places it in the oven. Delia, a famous chef and author of culinary textbooks, drops by. Fred produces the dish and says, "Would you like some apple pie?" Delia says, "Actually, that's not a pie. With a pie, the pastry must either completely enclose the filling, or the filling must be enclosed by a receptacle and have a pastry upper crust." Fred accepts the correction and goes on to ask what kind of dish he has produced. "A tart, I suppose," said Delia.

On Twin Earth, the word "pie" is used more widely, to cover various baked dishes, including, as it happens, the tart Fred has produced. When Twin Fred offers his friend Delia some apple pie, she merely accepts, having no dispute with Fred's description of the dish.

Now I'll let you in on some secrets: 'Earth' is England and 'Twin Earth' is the United States. The usage of "pie" does indeed vary across the Atlantic: witness American "pecan pie," "pumpkin pie," and "lemon meringue pie." The two versions of the story are both possible futures concerning Fred, who, although originally English, spends equal time in England and the United States.

What is the extension of the concept Fred expresses with "pie"? What does it depend on? The social externalist's position is that the extensions of one's concepts are partly determined by the views of the experts to whom one does and ought to defer. But here, two sets of experts seems equally suited. Fred would defer just as readily to American Delia as he would to English Delia. And it doesn't seem to be the case that one has more claim to Fred's allegiance than the other. Surely, the case could be fleshed out so that any non-arbitrary criteria of proper allegiance would apply equally to both candidates. Even if one could motivate the idea that it depends on what country Fred is in when he speaks, that would not help. Fred travels on the QE2 and spends much time mid-Atlantic. (Cases of this kind, in which the protagonist has allegiance to different communities, were offered in Loar 1987).

The social externalist is in a difficult position. How many concepts does Fred have? Two, because he defers to two sets of experts? None, because of his confused pattern of deference? Neither of these answers is plausible. Fred has exactly one concept that he expresses by "pie."

We can run a somewhat similar variant of Putnam's "elm" and "beech" case. The term "Milky Way" in England refers to a certain kind of rather light chocolate bar with a soft praliné like filling. In the United States, it refers to a type of chocolate bar with a soft nougat and caramel center, closely resembling what in England are called "Mars Bars." The closest American counterpart to what the English call "Milky Way" are in the United States called "Three Musketeers."

Fred is no expert in confections, and rarely eats chocolate. He believes that "Mars" and "Milky Way" denote different kinds of chocolate bars, but he does not know what distinguishes them. He is on the QE2, and fancies something sweet. "I think they have Milky Ways," says Delia. "Oh, fine, that's just what I want. Will you get me one?" What has Fred asked for? What does he want? As with the pie case, we can assume that neither English nor American usage dominates. We cannot motivate the idea that he wants an English Milky Way as opposed to an American one, nor an American Milky Way as opposed to an English one. Nor does the idea that he has two quite distinct desires glow with plausibility. Does he then have no concept at all? Of course he has a concept. He has a whole cluster of desires and beliefs that he expresses with the help of the expression "Milky Way." If he did not, it would be hard to explain his speaking as he does, his reaching into his pocket to give Delia some money, the fact that his actions are related to his being hungry, his wanting something sweet and so forth.

Note that it is easy to construct realistic cases. Consider for example, the transatlantic variations over "pavement," "garden," "football," "public school," "biscuit," "chicory," "endive," "chip," "fanny," "professor," "pants," "billion," and "turtle." Many of us may be in Fred's position (see Ludlow 1995).

The difficulty I have raised for the social externalist is this. The social externalist holds that provided one has a minimal competence with a term and is disposed suitably to defer to the right experts, then one can deploy the concept expressed by the term. But suppose there are two sets of appropriate experts. The externalist then faces a choice. He could hold that since there are two concepts made available by experts, and the subject has minimal competence with the word, and would defer to an expert of either of the two expert camps, he has two concepts. But this is not a good choice. First, it is obviously counterintuitive. Second, it ignores what the externalist should see as a confused pattern of deferential relations leading the subject to a confused state of mind. Third, it makes for poor psychological explanation: why attribute two concepts when there is only need for one?

Alternatively, the externalist could appeal to the abnormality and unsuitability of the social environment, and hold that the subject therefore lacks an expert concept. This leads to a further choice. Either the subject has no concept, or he has a different one from the experts. The former option is unacceptable, since we can and should attribute some concept to the protagonist, in order to explain his behavior and other cognitive processes.

And the latter option is problematic as well. If the subject does not have the expert concept, then what does he have? How do we find out? The problem now is not that these questions are unanswerable. Rather, it is that the social externalist now faces the same kind of questions as the internalist. For he must now give an account of how a subject can have a concept that does not depend upon his allegiance to any coherent group of experts.

The same problem arises in many other ways as well. Most obviously, it comes up in relation to the experts themselves. What is in the extension of the expert's concept of arthritis? Arthritis is inflammation of the joints. But what is the extension of the concepts *inflammation* and *joints*? It can't be experts all the way down. Something other than the opinion of experts must fix the extension of the experts' concepts.

Further, in real life, experts often fail to agree, and nothing determines which experts an individual does or should defer to. For example, in France "foie gras" has a legal meaning: the liver of any edible bird, including chicken. The important legal constraint on foie gras products is that they cannot include the livers of too many different types of bird. And it has a different culinary one: fresh or preserved liver of goose or duck (see the *Larousse Gastronomique*). The meaning of "chocolate" is at the time of writing in dispute between Brussels and chocolate manufacturers, and has a third dictionary meaning.

Dictionaries disagree too. Compare for example, the entries for "sapphire" in the *OED* (2nd edition) with the *Random House* (2nd edition). *OED* (entry b): "Min. Used as a general name for all the precious transparent varieties of native crystalline alumina, including the ruby." Random House: "Any gem variety of corundum other than the ruby, esp. one of the blue varieties." This disagreement is due neither to national differences nor to a difference between lay and scientific usage, because the *Oxford Concise Scientific Dictionary* (2nd edition) sides with the Random House, offering, "Any of the gem varieties of corundum except ruby, especially the blue variety, but other colors of sapphire include yellow, brown, green, pink, orange and purple."

Often, though, technical and ordinary dictionaries do offer substantially different definitions of the same term. For instance, "psychopath," as mentioned in chapter three, has a restricted meaning in many medical and psychological dictionaries and a much more general one in many non-technical dictionaries. Often, nothing will determine which of a plurality of dictionaries or experts has more claim to a lay speaker's allegiance.

Finally, dictionaries often contain multiple entries of a single polysemous word, specifying extensionally different but closely related meanings. Compare, for example, the *OED*'s entry b for "sapphire," given above, with its entry a: "A precious stone of a beautiful transparent blue." Again, nothing determines which of various entries in a dictionary deserves the allegiance of a reasonably competent but inexpert lay speaker.

It should be clear now that neither samples of natural kinds, nor experts, nor the conjunction of the two, can provide adequate constraints that can be fruitfully exploited by a completely general account of concepts. On any theory, many concepts do not have determinate extensions fixed by either. An internalist account of Oscar's *dwater* and Alf's *tharthritis* concepts is therefore not likely to face any special difficulties of its own. Rather, it is likely to face just the same difficulties that the externalist faces with concepts like "tree" or "fish," with the concepts of experts, with the concepts of inexpert transatlantic errants, with concepts associated with many polysemous words and so on.

5.3   PSYCHOLOGY FROM AN INTERNALIST PERSPECTIVE

Neologisms

I have claimed that if we are correctly, precisely and explicitly to describe Oscar and Alf, then we would do well to adopt neologisms to talk about their thoughts. This section should help alleviate any misgivings my claim might induce.

The first thing to note is that the proposal does not threaten us with widespread revision of current lay or scientific practice. There are two points to consider. The first is that the proposal has no ramifications for the correct, precise and explicit description of many concepts of many subjects. The second is that there may well not be any pressing need to give correct, precise and explicit descriptions of the minds of subjects whose concepts differ somewhat from those of the describer's. Let us consider these points in turn.

The need to adopt neologisms evidently only arises in cases where the ascriber lacks a ready word for a concept of the ascribee's. The argument of chapter two showed that Alf lacks the concept *arthritis*. Whatever concept it is that he expresses by that term, it is not the same as the experts'. However, if we consider cases in which subjects are expert in the use of the focal terms, the argument does not apply.

Putnam's twin experiment helped to show that Oscar does not have our concept of water. It does so because it exhibits some particular features. Twin Oscar's term "water" is conditioned by paradigm samples that do not fall under our term "water." So Twin Oscar's "water" expresses a different concept from ours, one with different extension conditions. If we now make the internalist assumption that Oscar and Twin Oscar have the same concepts with the same extension conditions, we must conclude that Oscar's concept differs from ours. Since we do not have a term to express Oscar's concept, we need to find a new one.

It is important to be aware that there are severe constraints on this sort of Putnamian thought experiment.

There are many Earth subjects and Earth terms that do not give rise to twin experiments meeting those constraints. Suppose that our Earth subject (call her "Eartha") is relatively expert in the use of the focal term, *T*, which extends over *K*s: she knows what dictionaries say about *T*, she has a largely correct theory about the nature of *K*s, she has had many and varied interactions with samples and so forth. And suppose further that *T* is not a natural kind term, that there is no underlying, unobservable property of a sort that it would take science to discover, which would be a good criterion for applications of *T*. Typically, in these cases it is very difficult to set up a plausible twin case of the standard sort. For it has to be that Eartha's twin ("Teartha") has her use of *T* conditioned by samples that do not fall in the extension of Eartha's *T*. Given Eartha's deep and detailed knowledge about the meaning of *T* and the real nature of *K*s, it is difficult to see how Teartha could exist in an environment that tied her usage to non-*K*s.

If we do manage to construct the case, what we find is that Teartha is subject to so many misapprehensions about her environment that it is natural and reasonable to hold that she has the very same concept that Eartha and we have: *T* in her mouth is true of *K*s and *K*s only, just as it is in ours. When she calls a non-*K* *T*, she speaks falsely. Let us look at a couple of examples.

Consider Eartha's concept *square*.[9] Eartha and Teartha both say "by definition, a square is an equilateral rectangle." They are both good at geometry, and could persuade you that they know what they are talking about: they know all about straight lines, sameness of length and angle and so on. Let us try to construct a Putnam style twin case. We have to suppose that Teartha has on many occasions applied "square" to non-squares. How are we to deal with the counterparts of those occasions when young

Eartha carefully measured the lines and angles of the squares in her exercise book, to make sure she was doing her geometry homework correctly; or the afternoon she spent cutting tiles for her kitchen floor, taking care to shape them into squares? How could Teartha go through parallel interactions with non-squares? Let us suppose that somehow this is possible. Tricks of the light and prestidigitator's skulduggery have somehow made it so. It is surely reasonable to think that in such a case, Teartha has just got it wrong. She means *square* by "square," and mistakenly thinks that it is squares that she is involved with. If she were apprised of the situation she would say: "It was just a bunch of tricks. I thought they were squares but they were not."

Consider Eartha's concept *fish*. It is the same story. Both Twins insist that to be a fish is to be a vertebrate, finny, cold-blooded animal that has gills throughout its life. Both can explain how 'fish' breed, breathe, swim and so on. Moreover, Eartha has often been fishing, and has cooked, filleted and eaten many fish. Again, if we try to set up an environment in which Teartha systematically calls non-fish "fish," we will find it reasonable to think that she is the subject of systematic mistakes.

I think you'll find that the same results arise for vast numbers of concepts. Kindly consider the concepts expressed by the following nouns (in at least one of their meanings): "arthritis," "beacon," "chapel," "doubt," "election," "finish," "gallop," "hunt," "intuition," "juggler," "kitchen," "language," "machine," "native," "omission," "principle," "question," "rain," "system," "teacher," "upholstery," "virtue," "whim," "xylophone," "youth," "zone." In each case, consider an Eartha who knows the definition of the term and has associated theoretical knowledge and practical experience (in cases where these are options). In each case, if you try to set up a twin case modeled on Putnam's paradigm, you will find it very difficult. And if you can set up the case you will find it plausible enough that Teartha has the Earth concept.

I am not saying that the interpretations on which Teartha has the Earth concept are mandated simply by consideration of the examples. There may also be externalist interpretations that are not immediately counterintuitive or obviously incorrect. My point is simply that the interpretation I recommend is at least as reasonable. If that is right, then the general strategy of attributing to twins concepts with the same extension conditions can reasonably be deployed here without any need to invoke special terms.

Nevertheless, this book is committed to the idea that there is at least some conceptual variation. Oscar's "water" concept differs from ours, as does Alf's "arthritis" concept. In each case, the relevant subjects differ over what they take to be crucial criteria for application of the terms: we (I and most of you) think that something must be H_2O if it is to be in the extension of "water" and that an affliction must in the joints if it is to fall under "arthritis." Oscar has a different view of "water," and Alf of "arthritis." My arguments indicate that these differences indicate differences of content. (Which is not to say that they cause or constitute such differences).

It may be that conceptual variation is widespread. Burge cases may be common in real life. It is easy enough to find subjects who are ignorant of definitive criteria for terms. In these cases, subjects and experts will associate different concepts with the same terms. And perhaps different, partially informed subjects will also associate different concepts with the same terms. If that is right, then psychologists who are expert in respect of the terms they use and who lack other words for the purpose, would need to introduce neologisms, if they were to describe the concepts of these subjects correctly, precisely and explicitly.

However, it is also possible that conceptual variation is the exception rather than the rule. It is possible, for example, that there is an innate, species wide "water" concept. This would of course be Oscar's concept, *dwater*, not our scientifically conditioned one. It is also possible that Alf's concept is widely shared. It may be a concept that anyone who has the standard sort of partial knowledge of the semantics of "arthritis" naturally homes in on. Perhaps then, a first idealization would be that for a large class of words there are two concepts abroad amongst those whose words they are: a largely innately determined, perhaps rather vague, lay concept that humans quickly and naturally attain in the right environments and, on the other hand, a technical, expert concept that would normally be explicitly and formally learned from a teacher or a book.[10]

Even if conceptual variation is rife, however, it does not follow that we are doomed to any manic introduction of neologisms. In ordinary practice, we can get by perfectly well without being precise and explicit. We can, for example, say "Alf believes he has arthritis in his thigh," "arthritis" being our best shot and good enough for most purposes.

And widespread conceptual differences among individuals, although a good subject of study as a general phenomenon, need have little impact on the practice of scientific psychology. This is simply because scientific psychology does not, by and large, study the idiosyncrasies of particular individuals. Like most science, its concern is with the general, not the particular. Detailed studies of particular individuals would normally be carried out either when an individual is of special interest (due, for example, to an unusual pathology or a special talent) or as a case study in which the individual is representative of a larger population.

Indeed, if the majority of a normal individual's concepts are idiosyncratic, then particular idiosyncrasies are not likely to be of scientific interest. It is the general facts about the nature, extent and causes of individual differences that would be of interest. On the whole, particular concepts will be of interest to the extent that they are not idiosyncratic, but common across the whole species or some significant population within it.

By and large, then, my proposals should have little impact on the way scientific psychologists describe their subjects.

Neologisms are likely to be of use for psychologists studying subjects that differ in some important and general way from the academics themselves, such as children, subjects from non-scientific cultures, or people (e.g., scientists and philosophers) from history. Here, the relevant generalizations will be over groups, not individuals, and the neologisms will signal important differences between the subject group's concepts and those of the scientists who study them. In these cases, the use of neologisms seems to be a rather good idea. The point warrants a little discussion.

The use of neologisms already has some currency. Some psychologists (e.g., Perner 1991) already use the term "prelief" to denote a concept of three-year olds', which they see as an undifferentiated concept that, on maturing, differentiates into our concepts of belief and pretence.

According to many developmental psychologists, small children have many concepts that we lack. For example, as Piaget claimed some time ago, children have a single concept that differentiates into our concepts of weight and density. It is not simply that they have the concept of weight and later acquire the concept of density. Rather, the ancestor concept runs together features of both in a way that makes it implausible that it is either the concept *weight* or the concept *density*. (This is argued in Carey 1985). We do not have this undifferentiated concept, nor do we have a word to express or denote it.

Carey (1985) also argues at some length that children lack the adult concept of living thing. The concept that children express by "alive" contrasts not with inanimate, but with "dead." Moreover, this contrasting pair of concepts itself runs together the contrasts between "real" and "imaginary," "existent" and "nonexistent," and "functional" and "broken." For example, children tend to say such things as, "A button is alive because it can fasten a blouse," "A table is alive because you can see it." And if you ask children to list things that are not alive, they tend to list only the dead: George Washington, ghosts, dinosaurs.

Carey argues precisely that the child's "alive" concept has a different extension from the adult's. The concept does not extend over plants. The reason for this is not merely that children don't call plants "alive." Rather, it is that their "alive" concept has its place not in a theory of biology, but in a theory of psychology. It is intimately connected with their theory of persons and other minded beings. The functions of living—eat, sleep, and so on—are explained in social and psychological terms, not biological ones. When children learn that both animals and plants are alive (usually by the age of about ten), this involves a very large and significant restructuring of the way they represent things, the inductions they make, the kinds of explanations they offer and so on (Carey, 1985, 186–190).

It is true that there is something of an internalist slant to Carey's work.[11] But even the die-hard externalist would find it very hard to argue that there are no cases of significant conceptual change between infancy and adulthood.

The history of science provides similar examples. Thus, before the time of Black, scientists had not distinguished heat from temperature. The Experimenters of the Florentine Academy, for example, had a single concept that ran together elements of both heat and temperature: sometimes they treated the quantity as intensive, sometimes as extensive. Here, it certainly seems right to attribute to the early scientists an incoherent concept that we lack. There is no particular reason why externalists should wish to deny this.

In all these cases we would do well to adopt neologisms. We could say, for example, that small children believe that they are 'shalive', not that they are alive, and that the Experimenters of the Academy believed that 'hemperature' had both strength and intensity. It is true that coining neologisms is not the standard practice of psychologists and historians of science, Perner notwithstanding. Carey, for example, uses the term "*heat*," in italics, for the concept of the experimenters of the Academy. But she clearly intends this term not to express our concept of heat. In effect, she adopts the policy that I claimed was standard in everyday propositional attitude reporting. She uses a word of her vocabulary in opaque contexts in a nonstandard way, with a nonstandard meaning. The context of discussion makes it clear that she is doing this, and provides us with enough clues to get something of a handle on the experimenters' concept. There is nothing wrong with this practice. But it would be clearer and less prone to induce confusion if one adopted two terms rather than using one ambiguously.

Methodology

In this concluding subsection I want briefly to sketch how psychology might look from an internalist perspective. None of what follows is supposed to be an argument for internalism. It is supposed merely to present a sketch of psychology that is compatible with internalism and independently plausible.

Jerry Fodor (1987, chap. 1) elegantly captured a fundamental feature of psychology. Psychological states have both representational properties (content properties) and causal powers. These mirror one another. One can predict the causal powers of a psychological state from its content and type (belief, desire, etc.). Psychology, both folk and scientific, specifies principles that allow us to exploit this correspondence.

Two simple cases will illustrate. Beth believes that either Brazil or Argentina will win the soccer world cup. When Argentina lose in the semi finals, she comes to believe that they will not win the world cup. Now we can predict who she thinks will win. When we describe her beliefs, we do so in terms of their contents. And, on the basis of this, we can predict the causal relations among them.

Beth believes that Brazil will win the world cup. She also believes that anybody who bets a hundred pounds on the winning team stands to win a lot of money. Beth wishes to win a lot of money. She has no reason not to place a bet on Brazil and believes she is in a position to do so. We predict that she will try to place a bet on Brazil.

Beth's desires and beliefs tend to cause what they tend to rationalize.[12] This coordination of causality and rationalization lies at the heart of psychology. And it offers us an obvious heuristic for ascribing contents: charity. We can get a lot right, if we attribute psychological states to subjects that would render their behavior rational: if the individual were subject to the states we posit, then she would behave as she does, given that the states tend to cause what they tend to rationalize.

Charity is certainly not the be all and end all of psychology. People often fail to draw logical conclusions from what they already believe, even when these consequences would matter to them. And people often fail to do what it would be rational for them to do, given what they believe and what they want. There are many different reasons for these limitations. There are heavy constraints of time and processing resources, such as memory. Moreover, people don't always use valid principles of reasoning when they do make inferences. Sometimes they use fallacious ones. Sometimes they fail to use valid ones, even if they are aware of them. And sometimes people are simply overwhelmed by the power of their desires, felt needs and addictions and go right ahead and do what they know to be the wrong thing.

For those and many other reasons, it is not always the case that the most charitable interpretation of a subject is the correct one. Nevertheless, charitable interpretation is a good basic heuristic, and one that plays a central role in common sense psychology.

Charity also applies to causal relations between the environment and psychological states. We get a lot of practically indispensable evidence about people's psychological states by looking at how they are situated in their environments. If we see a person running rapidly away from a manifestly enraged dog, we might hypothesize that he is afraid of it. But we have to be careful how we see the role of the environment. For it is not even approximately true that people believe all and only truths about the world they inhabit. Rather, we know that humans typically have five senses, and we know a fair amount about the ranges and limitations of these senses. For example, we know that in good light, a normal subject with her eyes open will form largely accurate visual representations of the shapes, sizes and locations of middle-sized objects in front of her. But we also know that she can't see objects that are in the dark, or too small, or too far away. And we know some of the conditions under which she will form false visual representations. We are thus in a position to make decent hypotheses about cognitive states that are caused by the environment.

Even with all its limitations taken into account and catered for, the charitable strategies of common sense psychology are only heuristics. They provide only a rough and preliminary guide to true ascriptions. This is in part because charitable interpretation doesn't require us to look inside the heads of our subjects, which is where the crucial causal activity takes place. The ideal common sense psychologist could be fooled by a giant look-up table or a good actor (see Block 1981 for discussion). Nevertheless, it is pretty good heuristic and it allows us to formulate and partially confirm psychological attributions.

Now there is nothing in the basic principles of common sense psychology, in the charitable principles of interpretation that it deploys, that entails externalism. It is perfectly possible that the best interpretations of a subject would not distinguish twins. So a first shot at an answer to the question of how we could find what the narrow contents (i.e., the contents) of a subject's concepts are, is that we can do so by the normal methods.

If Oscar were brought forward in time, we could begin to find out about his concept of dwater simply by using the normal methods. For example, we would look at what kinds of samples he is willing to call "water." And we would, of course, consider not just actual causes, but counterfactual ones too. What would cause him to assent to "water"? Since XYZ would cause him so to assent, we have a little evidence already that it is in the extension of the concept.

Commonsense interpretation tells us a great deal about what is not in the extension of a concept. If Oscar, in full command of the situation and apprised of all relevant facts, insists that a sample of oil is not what he calls "water," then this is good evidence that dwater does not include oil. We could pursue the matter. We might ask him why this sample isn't dwater (by saying, "Why isn't this water?" in Oscar's language). He might examine the sample and say, "Water is typically colorless, odorless, tasteless and not at all viscous: this sample is black, smelly, tastes awful and it is slightly viscous. I'd say this is oil and definitely not water." Excellent evidence that oil is not in the extension of Oscar's concept of dwater, wouldn't you say?

Consider also a systematically misleading twin Earth. There, alien scientists have engineered the environment so that most of what appears like water is really oil. It is actually black and smelly, but using complex tricks of the light and air the scientists make it seem like water. Misled, Oscar calls this oil "water." Here, ordinary principles of interpretation again indicate that we should not include oil in the extension of "dwater." For Oscar has a whole network of further beliefs about dwater and about the environment that undermine this interpretation. He believes, for example, that dwater is typically transparent, odorless, tasteless and fluid. He also believes that the samples he sees have all these properties. And it is in part because he believes these things that he calls the samples "water." He also believes that he is not in the clutches of alien scientists and not subject to systematic visual illusions. And, were the true situation revealed, he would withdraw his previous statements about what was correctly called "water."

These sorts of considerations strongly suggest that it is better to think that a brain in a vat is systematically deluded than that it has many true thoughts about pulses of electricity. Charity does not automatically mandate interpretations that rule out large scale falsity. For it must take into account the network of important background beliefs that support the subject's judgements. When such beliefs are revised, the subject is prone to sweeping changes of mind. If a brain were removed from its vat, placed in a body and informed of its situation, it might well feel inclined to say such things as: "Well, I thought I was seeing a world of tables and chairs, grass and trees, cats and dogs. But I was wrong."

So we could make considerable progress towards finding out what is in the extension of "dwater" just by using the standard means. *Dwater* is the concept that fits best into our best overall interpretation of the subject's behavior, taking into account all the relevant counter-factuals about what they would say and do under various circumstances.

There is no reason to suppose that the best overall interpretation would distinguish twins, and some reason to suppose that it would not. For the relevant counterfactuals about the subjects' behavior would probably be the same for twins. The beginning of internalist psychology is thus not some mysterious theory of the future. It is simply normal folk psychology applied with due care and attention.

It might be objected that the normal folk psychological judgements about Oscar and Alf appear to be externalist, so it does not seem that folk psychology has the internalist bent I am claiming for it. I agree that folk psychology is not unequivocally internalist and has some externalist strands. But folk psychology does not have nearly as much of an externalist slant as is sometimes made out. For one thing, many folk psychologists don't have the Putnamian and Burgean intuitions. And, as I argued in chapter two, it may be that in folk psychological attributions we sometimes use words with altered senses and extensions. So it may be that in saying "Alf believes he has arthritis in his thigh," we are not attributing the concept of arthritis, but gesturing at the concept of tharthritis. Further, the main arguments for internalism offered in chapters one and two, as well as in this chapter, are themselves based on central features of folk psychology.

The important point is that the basic principles of folk psychology can be applied in a way that is compatible with internalism. There is nothing about the basic apparatus of charitable interpretation, exploiting the parallel between causal and rational properties, that dictates that it would yield externalist interpretations, ones that would distinguish twins. Rather we should expect the reverse.

Folk psychology is just the beginning. To learn more about concepts, we would have to proceed to science. But scientific psychology exploits the same basic apparatus as do the folk. Cognitive science with its computational models and cognitive psychology and psycholinguistics with their attributions of complex tacit theories, all rely on something like the basic principle of charity. They ascribe representations and bodies of knowledge on the assumption that subjects, modules and cognitive systems behave in ways that make reasonable sense. In this way, they exploit the parallel between causal and representational properties of psychological states, events, and processes.[13]

Scientific psychology of course revises and extends the folk apparatus in many different ways, bringing in many further constraints on attributions of content, constraints from acquisition, deficits, computational models, neural scans and so on and on. To be sure that none of these constraints brings in externalist principles of attribution would require examination of the various different branches of psychology. This would be a most worthwhile pursuit.[14] But prima facie there is not the slightest reason to suppose that any of them do. Scientific psychology as it is actually practiced appears to be perfectly compatible with internalism. So the right way to find out about narrow contents is just the right way to find out about cognitive content generally: do psychology.

Notes

Chapter Two

1
Thomas Kuhn, personal communication. The point is also made in Fodor 1994.

2
Noam Chomsky holds this, as do some other prominent linguists.

3
Fodor (1994) tentatively endorses this version.

4
To be precise, it allows that coextensive atomic concepts can differ in content. The point is that the difference in content cannot be put down to a difference in the extension of component concepts. Water and H_2O might be held to differ in content in that the latter has a component concept of hydrogen, say, while the former does not. The view at issue allows for distinctions of content that do not involve any such differences in extension.

5
I think Burge (1982), McDowell (1984, 1986), Evans (1982), and Wiggins (1980) hold something along those lines.

6
This is a strong externalist constraint. It is possible to formulate a weaker thesis of world dependence, one that does not strictly require actual interactions between thinkers and instances of the kind. I will discuss this weaker version toward the end of this chapter.

7
See Williamson (1998) for related discussion.

8

See Boghossian 1997 for a related challenge to the externalist account of Dry Earth. See McLaughlin and Tye, forthcoming, for an attempted response.

9

Davidson (1992, 1994), for example, holds that a being enjoys states with cognitive content only if it relates appropriately both to the environment and to at least one other sentient creature: there must be at least some shared responses to shared stimuli. My main worry about Davidson's overall picture is that it requires too much sophistication on the part of the subject: neonates and animals are left out of account.

Chapter Three

1

Kaplan 1989, 602. See Mercier 1994 for discussion.

2

Loar (1987) runs essentially this argument using "arthritis" and the French "arthrite" for w and w'. See also Kimbrough 1991 for discussion.

3

The idea that second order beliefs can motivate some kind of semantic distinction among any candidate synonyms goes back to Mates 1950. The relevance of Mates cases to the present argument was pointed out to me by Mark Sainsbury and Michael Tye.

4

Internalism in the first sense does not entail internalism in the second, some externalist rhetoric notwithstanding.

5

The basic idea behind this sort of account is due to Davidson (1969). Davidson sees the semantic objects of belief reports as token utterances. I prefer to see them as sentences: abstract, syntactically structured objects with semantic properties. The latter sort of account has been developed in considerable detail, and works extremely well, both in respect of formal tractability and coverage of data. See Higginbotham 1986, Larson and Ludlow 1993, Larson and Segal 1995, chap. 11, Segal 1996b.

6

I take the content of the second order belief to be meta-linguistic as well. See Segal 1998 for details.

7
As mentioned above, we are not conscious of the semantic mechanisms we use in belief reporting, so there is no reason why what I am saying should be obvious to any of us.

Chapter Four

1
Compare: an object is fragile iff it has the second order property of having some first order property that causes it to break easily on impact. The first order properties can vary from case to case: one cup might be fragile because it is very thin, another because of irregularities at the boundaries of the crystals of which it is composed. In the psychological case, the first order states might have to do with patterns of connectivity among neurons in your brain or the organization of silicon chips in an android's.

2
Actually we have an account of narrow psychological states, rather than narrow content, since the Ramsey sentence of T doesn't specify what is in common among different types of state with same narrow content e.g. desires, beliefs and fears that p. These all have the same content, but different functional roles. There are different ways one could develop the proposal to deal with this. One possibility would be to take T to be a theory of mental representations, rather than psychological states. The idea would be that desiring, believing and fearing that p involve different relations to the same type of mental representation. T would specify the causal role of a representation when it features as the object of various kinds of states, and its narrow content could be identified with this causal role. My objections to narrow functionalism should apply, however this issue is dealt with.

3
I do not know if anybody has explicitly articulated and endorsed $DRNF_1$ in print, although it is at least suggested by Braddon-Mitchell and Jackson (1996, 220).

In any event, various philosophers I have talked to appear to have something like it in mind.

4
See Segal 1996a for a discussion of different varieties of modularity.

5
See Block 1980, 291–293, for a similar objection.

6
The functionalist could posit a fundamental level of categorical

properties. One worry about that would be that properties of the very small look like better candidates for functional reduction than some more macroscopic properties.

7
Listening to Nancy Cartwright led me to this view, which I think is hers.

8
The old problem is often called "Hume's problem." Crudely put it is, What distinguishes mere conjunction of events from causation? John Carriero informed me that there was little originality in Hume's formulation of the problem. Analogous points are made by Leibniz and the Port Royal Logic, and the problem was probably well known in Hume's time.

9
This is basically the Language of Thought model of cognition. See Fodor 1975, 1988, and Field 1978.

10
The locus classicus of the idea that conceptual role could be one factor in a two-factor account of content is Field 1977. See also McGinn 1982 for discussion.

11
The argument was first made in Noonan 1986 and is spelled out in detail in Segal 1989b.

12
They are not, of course, narrow. But it is intrinsicness that matters.

13
The term and the basic idea are due to Kaplan (see, e.g., 1977). Kaplan's account is designed to apply to language, not thought, so he talks of contexts of utterance. Note also that Kaplan makes a three-fold distinction among character, what he calls "content" and extension. I will discuss only a simplified version of the account.

14
The examples all involve complex concepts. So it is possible that cognitive content is a function of character and compositional structure. Maybe it would be worth pursuing this suggestion. It would require, once again, an account of the vocabulary we would use to specify cognitive content, so constructed, and the way it could be used to frame psychological generalizations.

15
The idea is originally due to White (1982). For similar proposals,

see Chalmers 1996 and Jackson 1998. I think that all of these accounts suffer from the problem that different cognitive contents can generate the same character.

16
Indeed, I think it is even worse than that. For "called" itself appears to be a deferential concept. Neophyte philosophers sometimes accept correction from expert philosophers in respect of their views on what is called what and on what it is for something to be called something. (I recommend the reader to construct a Burge style thought experiment, parallel to the original "arthritis" case, for "called"). Now it seems impossible even to formulate the characters: "thing standing in relation R to 'disease' standing in relation R to 'arthritis' in my thing standing in relation R to 'the home environment'"? And is "relation" deferential too?

17
The account will not work for those sympathetic to social externalism, since, as mentioned above, on that view, all or nearly all concepts have wide contents. We are then not left with any narrow concepts to provide the descriptive material.

Chapter Five

1
See Kripke 1980 for a similar story.

2
He now prefers to say that if one group is right, then so is the other (LaPorte, p.c.).

3
In fact, the chemical terms "H" and "H_2O" are ambiguous. Deuterium (D) and tritium (T, 1 proton, 2 neutrons) are hydrogen isotopes. In chemical classification, the isotopes of an element are varieties of that element. So D and T are varieties of hydrogen. "H" can be used either to denote hydrogen, hence to extend over deuterium and tritium, or it can be used in contrast to "D" and "T" to denote just normal hydrogen. "H_2O" is similarly ambiguous. The usage that is most consistent with chemical vernacular in general allows "H" and "H_2O" their wider meanings.

4
For detailed discussion of "whale," see Dupré (forthcoming). For more general discussion of folk and scientific biological taxonomies, see Dupré 1990. Our views on these matters are much the same.

5
I have borrowed some of the examples from LaPorte and dug out the others from the *Oxford Concise Science Dictionary* and the *Oxford English Dictionary*.

6

For attempts to deal with some of these cases from a roughly Putnamian standpoint, see Sterelny 1983, Devitt and Sterelny 1987, Miller 1992, and Brown 1998. None of the accounts deals with all of the cases. I think that they would all have trouble with, e.g., "Grant's zebra," "sapphire," and "water." I doubt that any account retaining basic Putnamian intuitions could deliver plausible results across a broad range of cases without getting other ones wrong. But that is for you to judge, dear reader.

7
In fact it is not clear that in ordinary contemporary English (as described in dictionaries) "water" is true of steam and ice, although it is in my idiolect. That is another case that could have gone either way (and may in fact have gone both ways).

8
David Papineau pointed out to me that someone still subject to Putnamian intuitions might adopt my proposal in a two-factor theory. Thus Oscar's concept might be understood along the lines of "dwater in my home environment," where "dwater" extends over the hydroid motley and the rest consists in a special kind of indexical that works off of appropriate four-dimensional and perhaps causal indices.

This sort of two factor theory avoids the problems discussed in the previous chapter. And it has some plausibility where terms are used by speakers with something like Putnam's model more or less explicitly in mind, so it is worth exploring. I doubt, however, that terms are often used in this way. So the theory would probably have limited application. It is not clear whether some version of this proposal would help preserve social externalist views about extension: "tharthritis in my home environment" will not work.

9
The example "square" is used to related effect by McGinn (1989), Segal (1991), and Lewis (1994). Lewis says, "You know the recipe for Twin Earth examples. You can follow it in these cases too. But what you get falls flat." That's the claim I am trying to flesh out here.

10
If holists like Block (1995) and Davidson (e.g., 1974, 1977) are

right, then it will rarely, if ever, be the case that two individuals share even one concept. If that is right, then most of our ascriptions of content are not correct, precise and explicit. Too bad. For general discussion of holism, see Fodor and LePore 1992.

11
See Patterson 1991 for an argument that Carey's account of conceptual development is internalist.

12
In fact, as I said in chapter three, I believe that representational properties are themselves casually efficacious: it is because a state represents what it does that it causes what it does. I also believe that casually efficacious properties must be intrinsic. This leads to a short argument for internalism. However, I can't offer any non-question-begging defense of the premises, so the argument is of little polemical value.

13
For a different view, see Chomsky 1995, forthcoming.

14
Rather, it is a worthwhile pursuit. There is a lot of literature on this topic, with arguments offered on both sides. See, for example, Patterson 1991 on developmental psychology; Chomsky 1995, forthcoming, and Mercier 1994 on psycholinguistics; Cummins 1983, Wilson 1995 and Segal 1997 on cognitive science. There has been in particular a detailed debate about David Marr's computational theory of vision, with Segal (1989c, 1991) and Butler (1996a, 1996b) arguing that it is internalist, and Burge (1986a, 1986b), Davies (1991, 1992), and Egan (1995) arguing that it is externalist.

References

Almog, J., Perry, J., and Wettstein, H. 1989. *Themes from Kaplan*. Oxford: Oxford University Press.

Baron-Cohen, S. 1995. *Mindblindness: An Essay on Autism and Theory of Mind*. Cambridge: MIT Press.

Barrett, J. L., and Keil, F. C. 1996. "Conceptualizing a Non-natural Entity: Anthropomorphism in God Concepts." *Cognitive Psychology* 31: 219–247.

Block, N. 1980. "Troubles with Functionalism." In N. Block, ed., *Readings in the Philosophy of Psychology*, vol. 1, pp. 268–305. Cambridge: Harvard University Press, 1980.

Block, N. 1981. "Psychologism and Behaviorism." *Philosophical Review* 90: 5–43.

Block, N. 1995. "An Argument for Holism." *Proceedings of the Aristotelian Society* 95: 151–169.

Boghossian, P. 1997. "What the Externalists Can Know A Priori." *Proceedings of the Aristotelian Society* 97: 161–175.

Boyer, P. 1994. "Cognitive Constraints on Cultural Representations: Natural Ontologies and Religious Ideas." In Hirschfield and Gelman 1994.

Brown, J. 1998. "Natural Kind Terms and Recognitional Capacities." *Mind* 107: 275–303.

Burge, T. 1973. "Reference and Proper Names." *Journal of Philosophy* 70: 425–439.

Burge, T. 1974. "Demonstrative Constructions, Reference, and Truth." *Journal of Philosophy* 71: 205–223.

Burge, T. 1977. "Belief De Re." *Journal of Philosophy* 74: 338–362.

Burge, T. 1979. "Individualism and the Mental." In P. French, T. Uehling, and H. Wettstein, eds., *Studies in Epistemology*, Midwest Studies in Philosophy, no. 4, pp. 73–121. Minneapolis: University of Minnesota Press.

Burge, T. 1982. "Other Bodies." In Woodfield 1982, pp. 97–120.

Burge, T. 1986a. "Individualism and Psychology." *Philosophical Review* 95: 3–45.

Burge, T. 1986b. "Intellectual Norms and the Foundation of Mind." *Journal of Philosophy* 83: 697–720.

Burge, T. 1989. "Wherein Is Language Social." In A. George, ed., *Reflections on Chomsky*, pp. 175–191. Oxford: Blackwell.

Butler, K. 1996a. "Individualism and Marr's Computational Theory of Vision." *Mind and Language* 11: 313–337.

Butler, K. 1996b. "Content, Computation, and Individualism in Vision Theory." *Analysis* 56: 146–154.

Carey, S. 1985. *Conceptual Change in Childhood*. Cambridge: MIT Press.

Cartwright, N. 1983. *How the Laws of Physics Lie*. Oxford: Clarendon Press.

Cartwright, N. 1994. "Fundamentalism vs. the Patchwork of Laws." *Proceedings of the Aristotelian Society* 94: 279–292.

Chalmers, D. J. 1996. *The Conscious Mind: In Search of a Fundamental Theory*. Oxford: Oxford University Press.

Chomsky, N. 1975. *Reflections on Language*. London: Temple Smith.

Chomsky, N. 1980. *Rules and Representations*. New York: Columbia University Press.

Chomsky, N. 1986. *Knowledge of Language*. New York: Praeger.

Chomsky, N. 1995. "Language and Nature." *Mind* 104: 1–61.

Chomsky, N. Forthcoming. "Internalist Explorations." In M. Hahn and B. Ramberg, eds., *Individualism and Skepticism*. Cambridge: MIT Press.

Crane, T., and Mellor, D. H. 1990. "There Is No Question of Physicalism." *Mind* 99: 185–206. Reprinted in Mellor 1991, pp. 82–103.

Cummins, R. 1983. *The Nature of Psychological Explanation*. Cambridge: MIT Press.

Davidson, D. 1963. "Actions, Reasons, and Causes." *Journal of Philosophy* 60: 685–700. Reprinted in Davidson 1980, pp. 3–20.

Davidson, D. 1969. "On Saying That." In D. Davidson and J. Hintikka, eds., *Words and Objections: Essays on the Work of W. V. O. Quine*, pp. 158–174. Dordrecht: Reidel. Reprinted in Davidson 1984, pp. 93–108.

Davidson, D. 1970. "Mental Events." In L. Foster and J. W. Swanson, eds., *Experience and Theory*. Amherst: University of Massachusetts Press, 1970. Reprinted in Davidson 1980, pp. 207–225.

Davidson, D. 1980. *Essays on Actions and Events*. Oxford: Oxford University Press.

Davidson, D. 1984. *Inquiries into Truth and Interpretation*. Oxford: Clarendon Press.

Davidson, D. 1987. "Knowing One's Own Mind." *Proceedings and Addresses of the American Philosophical Association*, pp. 441–458.

Davidson, D. 1992. "The Second Person." In P. French, T. Uehling, and H. Wettstein, eds., *The Wittgenstein Legacy*, Midwest Studies in Philosophy, no. 17, pp. 255–267. Notre Dame, Ind.: University of Notre Dame Press.

Davidson, D. 1994. "The Social Character of Language." In B. McGuinness and G. Oliveri, eds., *The Philosophy of Michael Dummett*. Dordrecht: Kluwer.

Davies, M. 1991. "Individualism and Perceptual Content." *Mind* 100: 461–84.

Davies, M. 1992. "Perceptual Content and Local Supervenience." *Proceedings of the Aristotelian Society* 92: 21–45.

Devitt, M., and Sterelny, K. 1987. *Language and Reality*. Oxford: Blackwell.

Dupré, J. 1990. "Natural Kinds and Biological Taxa." *Philosophical Review* 99: 66–99.

Dupré, J. Forthcoming. "Are Whales Fish?" In S. Atran and D. L. Medin, eds., *Folk Biology*. Cambridge: MIT Press.

Egan, F. 1995. "Computation and Content." *Philosophical Review* 104: 181–203.

Evans, G. 1982. *The Varieties of Reference*. Oxford: Oxford University Press.

Field, H. 1976. "Logic, Meaning, and Conceptual Role." *Journal of Philosophy* 74: 379–409.

Fodor, J. 1980. "Methodological Solipsism as a Research Strategy in Cognitive Psychology." *Behavioral and Brain Sciences* 3, no. 1: 63–72. Reprinted in Fodor 1981, pp. 225–253.

Fodor, J. 1981. *Representations: Philosophical Essays on the Foundations of Cognitive Science.* Cambridge: MIT Press.

Fodor, J. 1987. *Psychosemantics.* Cambridge: MIT Press.

Fodor, J. 1990a. "A Theory of Content, II." In Fodor 1990b, pp. 89–136.

Fodor, J. 1990b. *A Theory of Content and Other Essays.* Cambridge: MIT Press.

Fodor, J. 1994. *The Elm and the Expert.* Cambridge: MIT Press.

Fodor, J. 1998. *Concepts: Where Cognitive Science Went Wrong.* Oxford: Oxford University Press.

Fodor, J., and LePore, E. 1992. *Holism: A Shopper's Guide.* Oxford: Blackwell.

Gopnik, A., and Meltzoff, A. 1997. *Words, Thoughts, and Theories.* Cambridge: MIT Press.

Gopnik, A., and Wellman, H. 1992. "Why the Child's Theory of Mind Really Is a Theory." *Mind and Language* 7: 145–172.

Harman, G. 1987. "(Non-solipsistic) Conceptual Role Semantics." In E. LePore, ed., *New Directions in Semantics.* London: Academic Press.

Haugeland, J. 1978. "The Nature and Plausibility of Cognitivism." *Behavioral and Brain Sciences* 1: 215–226. Reprinted in Haugeland 1981, pp. 243–281.

Haugeland, J. 1981. *Mind Design.* Cambridge: MIT Press.

Higginbotham, J. 1986. "Linguistic Theory and Davidson's Program in Semantics." In E. LePore, ed., *Truth and Interpretation: Perspectives on the Philosophy of Donald Davidson*, pp. 29–48. Oxford: Basil Blackwell.

Hirschfield, L. A., and Gelman, S. A., eds. 1994. *Mapping the Mind: Domain Specificity in Cognition and Culture.* Cambridge: Cambridge University Press.

Jackson, F. 1998. *From Metaphysics to Ethics: A Defense of Conceptual Analysis.* Oxford: Clarendon.

Kaplan, D. 1977. "Demonstratives." In Almog, Perry, and Wettstein 1989, pp. 481–563.

Kaplan, D. 1978. "Dthat." In P. Cole, ed., *Pragmatics*. Syntax and Semantics, no. 9, pp. 221–243. New York: Academic Press.

Kaplan, D. 1989. "Afterthoughts." In Almog, Perry, and Wettstein 1989, pp. 565–614.

Kaplan, D. 1990a. "Thoughts on Demonstratives." In P. Yourgrau, ed., *Demonstratives*. Oxford: Oxford University Press.

Kaplan, D. 1990b. "Words." *Proceedings of the Aristotelian Society*, suppl. vol. 64: 93–119.

Keil, F. 1989. *Concepts, Kinds, and Cognitive Development*. Cambridge: MIT Press.

Kim, J. 1984. "Concepts of Supervenience." *Philosophy and Phenomenological Research* 45: 153–176. Reprinted in Kim 1993, pp. 53–78.

Kim, J. 1993. *Supervenience and Mind: Selected Philosophical Essays*. Cambridge: Cambridge University Press.

Kimbrough, S. 1991. "Anti-individualism and Fregeanism." *Philosophical Quarterly* 98, no. 193: 470–483.

Kripke, S. 1971. "Identity and Necessity." In M. Munitz, ed., *Identity and Individuation*, pp. 135–164. New York: New York University Press, 1971.

Kripke, S. 1979. "A Puzzle about Belief." In A. Margalit, ed., *Meaning and Use*, pp. 239–283. Dordrecht: Reidel. Reprinted in Salmon and Soames 1988, pp. 102–148.

Kripke, S. 1980. *Naming and Necessity*. Harvard: Harvard University Press. Originally published in D. Davidson and G. Harman, eds., *Semantics of Natural Language*, pp. 253–355. Dordrecht: Reidel, 1972.

LaPorte, J. 1996. "Chemical Kind Term Reference and the Discovery of Essence." *Noûs* 30, no. 1: 112–132.

Larson, R., and Ludlow, P. 1993. "Interpreted Logical Forms." *Synthese* 95: 305–355.

Larson, R., and Segal, G. 1995. *Knowledge of Meaning: An Introduction to Semantic Theory*. Cambridge: MIT Press.

Leslie, A. 1987. "Pretense and Representation: The Origins of Theory of Mind." *Psychological Review* 94: 412–426.

Leslie, A. 1991. "The Theory of Mind Impairment in Autism: Evidence for a Modular Mechanism in Development?" In A.

Whiten, ed., *Natural Theories of Mind: Evolution, Development, and Simulation of Everyday Mindreading*. Oxford: Blackwell.

Leslie, A. 1994. "Pretending and Believing: Issues in the Theory of ToMM." *Cognition* 50: 211–238.

Lewis, D. 1970. "How to Define Your Theoretical Terms." *Journal of Philosophy* 67: 427–446.

Lewis, D. 1972. "Psychophysical and Theoretical Identifications." *Australian Journal of Philosophy* 50: 249–258.

Lewis, D. 1983. "Extrinsic Properties." *Philosophical Studies* 44: 197–200.

Lewis, D. 1994. "Reduction of the Mind." In S. Guttenplan, ed., *A Companion to the Philosophy of Mind*, pp. 412–431. Oxford: Blackwell.

Loar, B. 1987. "Social Content and Psychological Content." In R. Grimm and D. Merrill, eds., *Contents of Thought: Proceedings of the 1985 Oberlin Colloquium in Philosophy*, pp. 99–139. Includes comments by Akeel Bilgrami, pp. 110–121.

Ludlow, P. 1995. "Externalism, Self-knowledge and the Prevalence of Slow Switching." *Analysis* 55: 45–49.

Lycan, W. 1987. *Consciousness*. Cambridge: MIT Press.

Marr, D. 1982. *Vision*. New York: W. H. Freeman and Co.

Mates, B. 1950. "Synonymity." *University of California Publications in Philosophy* 25: 201–226. Reprinted in L. Linsky, ed., *Semantics and the Philosophy of Language*, pp. 111–136. Urbana: University of Illinois Press.

McDowell, J. 1977. "The Sense and Reference of a Proper Name." *Mind* 86: 159–185. Reprinted in A. Moore, ed., *Meaning and Reference*, pp. 111–136. Oxford: Oxford University Press.

McDowell, J. 1984. "*De Re* Senses." In C. Wright, ed., *Frege: Tradition and Influence*, pp. 283–294. Oxford: Basil Blackwell.

McDowell, J. 1986. "Singular Thought and the Extent of Inner Space." In Pettit and McDowell 1986, pp. 137–168.

McGinn, C. 1982. "The Structure of Content." In Woodfield 1982, pp. 207–258.

McGinn, C. 1989. *Mental Content*. Oxford: Basil Blackwell.

McLaughlin, B., and Tye, M. Forthcoming. "Is Content-Externalism Compatible with Privileged Access?" *Philosophical Review*.

Mellor, D. H. 1991. *Matters of Metaphysics*. Cambridge: Cambridge University Press.

Mercier, A. 1994. "Consumerism and Language Acquisition." *Linguistics and Philosophy* 17: 499–519.

Miller, R. B. 1992. "A Purely Causal Solution to One of the *Qua* Problems." *Australasian Journal of Philosophy* 70: 425–434.

Millikan, R. 1984. *Language, Thought, and Other Biological Categories*. Cambridge: MIT Press.

Millikan, R. 1989. "Biosemantics." *Journal of Philosophy* 86: 281–297.

Noonan, H. 1986. "Russellian Thoughts and Methodological Solipsism." In J. Butterfield, ed., *Language, Mind, and Logic*, pp. 67–90. Cambridge: Cambridge University Press.

Patterson, S. 1991. "Individualism and Semantic Development." *Philosophy of Science* 58: 15–31.

Perner, J. 1991. *Understanding the Representational Mind*. Cambridge: MIT Press.

Perry, J. 1977. "Frege on Demonstratives." *Philosophical Review* 86: 474–497. Reprinted in P. Yourgrau, ed., *Demonstratives*, pp. 50–70. Oxford: Oxford University Press.

Perry, J. 1979. "The Problem of the Essential Indexical." *Noûs* 13: 3–21. Reprinted in Salmon and Soames 1988, pp. 83–101.

Pettit, P., and McDowell, J. 1986. *Subject, Thought, and Context*. Oxford: Clarendon Press.

Putnam, H. 1975a. "The Meaning of 'Meaning'." In K. Gunderson, ed., *Language, Mind, and Knowledge*, Minnesota Studies in the Philosophy of Science, no. 7. Reprinted in Putnam 1975b, pp. 215–271.

Putnam, H. 1975b. *Mind, Language, and Reality*. Vol. 2 of *Philosophical Papers*. Cambridge: Cambridge University Press.

Ramsey, F. P. 1931. "Theories." In F. P. Ramsey, *The Foundations of Mathematics*, pp. 212–236. London: Routledge and Kegan Paul.

Salmon, N. and Soames, S. 1988. *Propositions and Attitudes*. Oxford: Oxford University Press.

Searle, J. 1983. *Intentionality*. Cambridge: Cambridge University Press.

Segal, G. 1989a. "A Preference for Sense and Reference." *Journal of Philosophy* 86: 73–89.

Segal, G. 1989b. "The Return of the Individual." *Mind* 98: 39–57.

Segal, G. 1989c. "Seeing What Is Not There." *Philosophical Review* 98: 189–214.

Segal, G. 1991. "Defense of a Reasonable Individualism." *Mind* 100: 485–493.

Segal, G. 1996a. "The Modularity of Theory of Mind." In P. Carruthers and P. K. Smith, eds., *Theories of Theories of Mind*. Cambridge: Cambridge University Press.

Segal, G. 1996b. "Frege's Puzzle as Some Problems in Science." In *Rivista di Linguistica* 8: 375–389.

Segal, G. 1997. Review of Wilson, *Cartesian Psychology and Physical Minds: Individualism and the Sciences of Mind*. *British Journal for the Philosophy of Science* 48: 151–157.

Segal, G. 1998. "Representing Representations." In P. Carruthers and J. Boucher, eds., *Language and Thought: Interdisciplinary Themes*, pp. 146–162. Cambridge: Cambridge University Press.

Segal, G., and Sober, E. 1991. "The Causal Efficacy of Content." *Philosophical Studies* 63: 1–30.

Shoemaker, S. 1998. "Causal and Metaphysical Necessity." *Pacific Philosophical Quarterly* 79: 59–77.

Spelke, E. 1988. "Where Perceiving Ends and Thinking Begins: The Apprehension of Objects in Infancy." In A. Yonas, ed., *Minnesota Symposium on Child Psychology*, vol. 20. Hillsdale, N.J.: Erlbaum.

Sterelny, K. 1983. "Natural Kind Terms." *Pacific Philosophical Quarterly* 64: 110–125.

Wellman, H. 1990. *The Child's Theory of Mind*. Cambridge: MIT Press.

Wellman, H., and Estes, D. 1986. "Early Understanding of Mental Entities: A Re-examination of Childhood Realism." *Child Development* 57: 910–923.

White, S. 1982. "Partial Character and the Language of Thought." *Pacific Philosophical Quarterly* 63: 347–365.

Wiggins, D. 1980. *Sameness and Substance*. Oxford: Oxford University Press.

Wiggins, D. 1986. "On Singling Out an Object Determinately." In Pettit and McDowell 1986, pp. 169–176.

Williamson, T. 1998. "The Broadness of the Mental: Some Logical Considerations." In James Tomberlin, ed., *Language,*

Mind, and Ontology, Philosophical Perspectives, no. 12, pp. 389–410. Oxford and Boston: Blackwell.

Wilson, M. 1983. "Predicate Meets Property." *Philosophical Review* 91: 549–589.

Wilson, R. A. 1995. *Cartesian Psychology and Physical Minds: Individualism and the Sciences of Mind.* Cambridge: Cambridge University Press.

Woodfield, A. 1982. *Thought and Object.* Oxford: Clarendon Press.

Index